The X-Ray Universe

Wallace Tucker
Riccardo Giacconi

Harvard University Press

Cambridge, Massachusetts, and
London, England 1985

Library of Congress Cataloging in Publication Data

Tucker, Wallace.
    The x-ray universe.

    (Harvard books on astronomy)
    Bibliography: p.
    Includes index.
    1. X-ray astronomy.   I. Giacconi, Riccardo.   II. Title.
III. Series.
QB472.T83   1985   522'.686   84-15654
ISBN 0-674-96285-0 (alk. paper)

For Karen and Mirella

Preface

This book is a selective and personal history of x-ray astronomy. We have made no effort toward completeness, and important contributions from other groups have not been given the credit or the space they deserve. For this we apologize to our colleagues. We have taken this approach, which is unusual for the Harvard Books on Astronomy series, for several reasons. First of all, x-ray astronomy is no longer a new and exotic branch of astronomy. It is now part of the mainstream, and a comprehensive treatment of x-ray astronomy as a separate field is no longer feasible. Hence we have selected for discussion topics that are of historical interest and are representative of the broad range of problems that can be attacked with x-ray observations. Second, the nature of our involvement in x-ray astronomy would have made an impersonal treatment difficult if not impossible, even if we had wanted to attempt it. Finally, by giving a first-hand narrative we hope to recapture our experiences in the development of this new field of astronomy, the good as well as the bad, and to communicate to the reader the excitement of doing science in previously uncharted areas. We also hope to contribute perspective on how we believe science, big and small, can be done in the space age.

We are grateful to our colleagues at American Science and Engineering, at the Harvard-Smithsonian Center for Astrophysics, and at other institutions around the world; their work has made the development of x-ray astronomy possible and their friendship has made our participation in the field a pleasure. We wish to acknowledge in particular the seminal contributions of Bruno Rossi and Herbert Friedman. Others have been especially important to us through their contributions to specific projects in which we were involved and the intellectual interaction they provided over the years. These include George Clark, Paul Gorenstein, Herbert Gursky, Edwin Kellogg, Stephen Murray, Minoru Oda, Frank Paolini, Kenneth Pounds, Ethan Schreier, Harvey Tananbaum, Guiseppe Vaiana, Leon Van Speybroeck, and John Waters. Finally, we thank Karen Tucker and Mirella Giacconi. They have been involved in this project from the beginning, when each of us first entered the field. Their support and encouragement have been essential and their comments on the manuscript have been most valuable.

Contents

The X-Ray Universe

1

The X-Ray Universe

Kennedy Space Flight Center, Cape Canaveral, Florida, just before midnight, November 12, 1978. On the launch pad a seventy-foot Atlas-Centaur rocket glistened white in the moonlight, wisps of vapor curling around its base as the countdown proceeded. T minus 30 minutes and counting. In the grandstands several hundred people shivered in the cool Florida night. Scientists and engineers waited, their palms sweating despite the cold. T minus 25 minutes and counting. Wives and children were there to see what their husbands and fathers had been working for all those years. T minus 20 minutes and counting. Hold at T minus 20 minutes.

Riccardo Giacconi was in the blockhouse, manning one of the fourteen audio channels used to monitor the status of the rocket and the spacecraft perched on top of it. The messages coming across those channels were hair-raising. "Possible leak in the hydrogen line." "Anomalous voltage level." The scientists realized that they were, during these crucial moments, totally dependent on people they had never heard of. More than five hundred people, each of whom had to do his or her job well, or there would be a disaster.

One of the drawbacks of doing science in the space age is that you have to wait a lot. The projects are large, complex, and expensive, and they take years to complete. Most of the scientists and engineers in the grandstands had been working on this project for five years. The ones who had been involved in the early planning stages had been at it for ten years or more. Others had come and gone over the years, leaving the project before they could taste the fruits of their labor.

Inside the blockhouse at Cape Canaveral. (NASA)

The other side of the coin is that space science has a dramatic quality that few other professions can match. The hopes and anxieties of all the long years are focused into two unequivocal moments of truth: launch, when either the rocket works or it doesn't; and activation, when by remote control you turn the instruments on and see if they work.

This aspect of space science makes it easy to recognize major turning points in the development of a new field. In November 1978 those who waited and watched in the grandstands and in the blockhouses and at the operations control center at Goddard Space Flight Center knew they were about to witness such a turning point. Sitting atop the Atlas-Centaur was a spacecraft that housed the largest and most versatile x-ray telescope ever built. With its launch, a new field, one that had been born with the space age, would come into its own. It would be possible to pinpoint the positions of strong sources of x-rays thought to come from such strange objects as neutron stars, black holes, and quasars, and to make detailed x-ray images of exploded stars and giant clouds of intergalactic gas.

Visible light, by which we normally see the stars, is but a small part of the overall spectrum of electromagnetic radiation. In many circumstances electromagnetic radiation can be thought of as an electromagnetic wave. The distance between crests in a wave is called the wavelength, and the rate at which crests pass a given point is called the frequency. The different types of electromagnetic radiation have different wavelengths. The wavelengths of low-frequency radiowaves are several meters, those of visible-light waves a few hundred-thousandths of a centimeter, those of x-rays less than a millionth of a centimeter, and those of gamma rays even smaller.

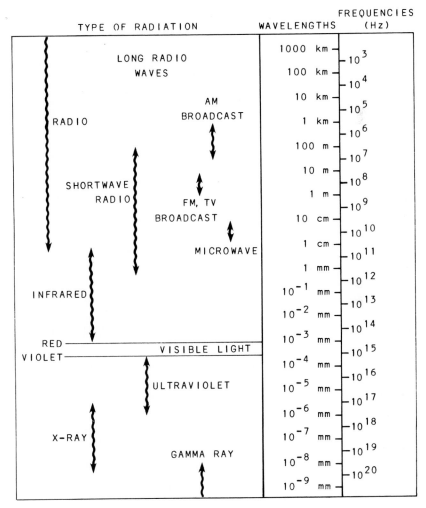

The electromagnetic spectrum. (From D. Goldsmith, *The Evolving Universe* (Menlo Park, Calif.: Benjamin Cummings, 1981))

In other situations electromagnetic radiation does not behave so much like a wave as like a stream of energy packets called photons. In general, if the stream of radiation contains a very large number of photons, then the wave description is more appropriate; if individual photons can be distinguished, then a description in terms of photons is more useful. In x-ray physics and astronomy, the photon description is almost always used. Wavelength is related to the energy carried by a photon: the shorter the wavelength, the higher the energy of the corresponding photons. For example, the wavelength of x-radiation is about a thousand times less than that of visible radiation, and the energy of an x-ray photon is about a thousand times more than that of visible-light photon.

The x-rays used in industry and medicine, as well as those detected by x-ray astronomers, must be produced in places containing high-energy particles. It is not surprising, then, that an x-ray image of the sky can look markedly different from an optical image. In essence, the x-ray images reveal the hot spots in the universe; regions where particles have been energized or raised to very high temperatures by strong magnetic fields or violent explosions or intense gravity. The temperatures are in many cases much higher than those inside the hottest stars. Where do such conditions exist? In an astonishing variety of places, ranging from the vast spaces between galaxies to the turbulent surfaces of stars and to the bizarre, collapsed worlds of neutron stars and black holes.

The x-ray universe remained unknown as well as invisible until the space age. X-rays interact strongly with the earth's atmosphere and are absorbed by it long before they reach the surface of the earth. To study cosmic x-rays, astronomers must use rockets to carry their detectors a hundred miles or more above the earth. Captured German V-2 rockets were used in the late 1940s in the first efforts to explore the x-ray universe, but not until the 1960s was the technology of rockets and x-ray detectors sufficiently developed to make possible a realistic attempt to detect x-rays from beyond the solar system. The search for cosmic x-rays grew out of an underlying belief that whenever a new way of observing nature was exploited, a rich variety of new phenomena would be discovered. In retrospect, the validity of this approach seems self-evident. At the time, however, it had few supporters. The proposal for the rocket flight that discovered the first x-ray star was rejected by NASA and was accepted by the Air Force only with the understanding that its primary goal was to look for x-rays from the moon.

2

The Sensible World

"Our subject is the sensible world, not a world on paper." Galileo put these words in the mouth of his character Salviatus more than 350 years ago. This contrast—between the "paper world" of natural philosophers in the Aristotelian tradition and the "sensible world of observation"—was at the core of the scientific revolution of the seventeenth century.

Since that time the views of the Galilean-Newtonian school have prevailed: observations and experiments have been considered essential building blocks of theoretical structures. Indeed, at the end of the nineteenth century the pendulum had swung so far toward the observational-experimental approach that self-styled disciples of Galileo and Newton, misunderstanding their legacy, denied the validity of pure theory in solving problems in the real world. They even actively persecuted theoretical investigators. A famous disagreement between Lord Kelvin and Ludwig Boltzmann demonstrated in a tragic way that philosophical beliefs about method in science, as opposed to the actual way science is done, can have a profound impact on the conditions under which scientific research is carried out.

Lord Kelvin took the (by then) conservative Galilean-Newtonian point of view: "The true naturalist knows that the essence of science consists of inferring antecedent conditions, and anticipating future evolutions from phenonema which have actually come under observation." On the other side of the debate, Boltzmann propounded pure theory as a way of understanding the physical world: "The more abstract the theoretical investigation, the more powerful it becomes; theory conquers the world."

Boltzmann was challenged by Kelvin and Peter Tait in the field of applied mathematics. He was isolated among the atomists, who saw in his theories "inconvenient excesses of generalization and abstraction," and he was attacked on philosophical grounds by Ernst Mach and Wilhelm Ostwald. At the 1895 Congress of Lubeck, he saw his view of the world and his scientific theories rejected by majority vote. This incident revealed more about the tenor of the times and the character of Boltzmann's opponents than about the veracity of his ideas, but it affected him profoundly. Although later generations vindicated him and recognized that he had both laid the foundation and built the structure of much of statistical mechanics, in his own lifetime he was hounded by his academic colleagues. Finally, ill and depressed, he committed suicide in 1906.

Today this intolerance and doctrinaire certainty that only one approach to knowledge of the physical world is valid is no more than a musty skeleton in the closets of science. The need for different approaches to scientific research is hardly a matter of debate. A conflict between the observational and theoretical approaches to the natural sciences no longer makes any intellectual sense. Modern research in the history of science has shown clearly that scientists do not carry out their work in a straightforward logical procession in keeping with the steps of the scientific method as outlined by philosophers, but rather that they take a much more complex and tortuous path, influenced by a multitude of conditions that includes the society around them as well as their own intellectual baggage. Technologically advanced nations spend enormous sums of public money every year to support experimental investigations in all scientific disciplines. At the same time, the very powerful and abstract theories of Bernhard Riemann, Paul Dirac, Wolfgang Pauli, and Albert Einstein dominate our conception of the universe.

If all is well, why do we wave the Galilean flag? Because we share the concern of many scientists that experimental investigations that are truly new and can lead to great discoveries are being hampered by the structured approach to the conduct of science. This approach appears to result from the bureaucratic control of federal funding for research. It is difficult to say whether such structure has been adopted by the bureaucracies more from need or from preference. But it seems to us clear that it rests on a diffused and simplistic perception of how science is done, and that it is aided by a widening schism between theorists and experimentalists. This schism does not result from a reasoned difference in philosophy, but rather is the effect of misunderstandings brought about by social customs and by increased specialization.

There is a generally accepted view that science proceeds according to the scheme outlined by Thomas Kuhn in his book *Scientific Revolutions*. We believe this view has had a negative impact on decisionmaking in science, especially through its influence on the bureaucracies that administer funds for scientific research in the United States. According to Kuhn a scientist approaches a problem with a mental picture developed over years of observing, hypothesizing, theorizing, and testing. Kuhn calls this mental picture, which the scientist shares with his colleagues, a "paradigm." He asserts that the paradigms accepted by scientists at any given time guide them in their research and determine how they will perceive the world.

The existence of shared paradigms in a science denotes its maturity. It is only in the early stages of development of a new field that its practitioners fumble along without any conceptual framework. Kuhn says that a mature field of science exhibits alternating periods of "normal science" and "scientific revolution." In times of normal science, paradigms appear to work well, explaining new observations as they are obtained, and the observations tend to extend and consolidate paradigms. From time to time, however, observations are obtained that do not fit these neat pictures. Eventually, but only after much searching for a way out, the old paradigm is abandoned in favor of new ones. Such periods are Kuhn's scientific revolutions.

Many historians of science consider Kuhn's approach only a rationalization after the fact of the way science is really done. Notwithstanding this criticism, it is easy to see that the neat division into periods of normal science and scientific revolution could have great appeal to a government official pressured to make difficult decisions and to justify these decisions simply and quickly to his superiors. To decide what should be done next in a field of science, the official can simply form a committee of eminent scientists, typically but not necessarily theorists, the "keepers of the paradigms," and ask them what should be the next step in a logical sequence of extending the validity of the paradigm.

But a paradigm is only a model on paper of the real world; it is not the real world. Outside the region illuminated by paradigms lies a whole world to be known. A discovery is by definition unpredictable from the limited point of view of previous knowledge. What then is to guide the scientist who wishes to embark on a voyage of discovery? Intuition, imagination, speculation, aesthetic considerations, a critical view of existing paradigms, intellectual arrogance, self-confidence, an almost childlike and mystical desire to know, a deeply held belief in the richness of nature. It would be difficult for a committee to defend its recommendations on the basis of reasons like these, yet often just such reasons motivate the leap of imagination and sustain the great effort that the would-be discoverer must dedicate to the pursuit of his vision.

The freedom to pursue an independent path has been more easily available to the theorists, whose traditional requirements have been pencil and paper (though nowadays they, too, often need considerable funding to pay for computer calculations), than to the experimentalists, who often need elaborate and expensive tools to carry out their work. This is certainly the case in elementary particle physics, where access to costly time at accelerators has to be carefully parcelled out, and in many areas of astronomy, where the proposals for use of the oversubscribed national observatories must go through a severe review procedure. The stifling of the independent spirit is an inevitable byproduct of these rigorous selection processes. This is of concern in any area of science where public funds are necessary to carry out the investigations, especially for those major undertakings which go under the label of "big science."

Several authors have claimed that the outstanding discoveries stem from relatively small enterprises carried out by a few individuals. For example, pulsars were discovered by a small group with a limited budget at Cambridge University, rather than by a big-science team. We believe that the problem has less to do with big versus small than with the restrictions that are inevitably imposed on big-science teams. It is extremely difficult for a big-science team to pursue the kinds of independent, innovative lines of research that often lead to major discoveries. If large amounts of public funds are to be spent, the expenditure must be justified, and for justification the government officials will turn to the keepers of the paradigms.

The schism in science today is not so much between the observer or experimentalist and the theorist as between the keepers of the paradigms and those who would follow a course that would take them outside the paradigms. A related problem is that the keepers of the paradigms often consider the observers or experimentalists to be merely technicians. This view ignores the personal and unique contributions that experimentalists can make in their search for new and unsuspected glimpses of nature's richness. These contributions do not come primarily from the execution of the task, however complex, but rather from seeing the problem in a new frame of reference and from posing questions that can be answered, so that we can probe in a new way the natural phenomena around us.

X-ray astronomy developed because creative scientists ignored the predictions of theorists that a search for cosmic sources of x-rays would be futile or of little interest, and invented ways to use this region of the electromagnetic spectrum to observe previously undetectable aspects of the universe. The result was the opening up of one of the most exciting fields of modern astrophysics.

Of course, x-ray astronomy did not begin in a vacuum, nor has it blossomed in isolation from the rest of astronomy. It emerged as part of a revolution that has occurred throughout the field of astronomy. This revolution began in the 1950s, when newly developed techniques for detecting radio waves were applied to the search for sources of cosmic radio waves. A series of spectacular discoveries came in rapid succession: first the discovery that the supernova explosion of a star generates a vast cloud of gas and high-energy particles; then the discoveries of radio galaxies and quasars, which led to the conclusion that explosive events that release a million times as much energy as a supernova explosion take place on a galactic scale; and finally the discovery of the microwave background radiation, which provided strong evidence that an explosive event, the Big Bang, encompassed and very likely gave birth to the Universe. By the end of the 1960s it was clear that violent events and high-energy processes play crucial roles in the universe.

As x-ray astronomy developed, it became apparent to many scientists that observations at x-ray wavelengths are uniquely suited to studying certain features of the high-energy universe. Neutron stars, the stellar remnants of supernova explosions, radiate large amounts of x-rays, as does the hot gas behind shock waves produced by supernova explosions. The central cores of galaxies, which may contain supermassive black holes, are also strong sources of x-radiation. On a much larger scale, regions of space where thousands of galaxies have clustered together are filled with vast clouds of hot gas that glows in x-rays.

The brief history of x-ray astronomy in many ways epitomizes modern science. It has the tension between experiment or observation and theory. It has the competition between groups to become the first to discover some new phenomenon, or for the opportunity to use a spacecraft; it has failures, misinterpretations of data, ventures down blind alleys, and also those golden moments when everything works, when everything comes together, theory and experiment, and we understand the universe a little better than we did before.

3

Precursors

In early November 1895 Wilhelm Roentgen did something many physicists before him had done. He went into his laboratory and turned on the power to a Crookes tube, a glass tube into which two conducting plates had been sealed in a vacuum. The Crookes tube was the forerunner of the cathode ray tubes (CRTs) used today in television sets and home computers. The cathode is the positively charged plate or electrode, and the cathode rays are electrons that carry a charge across the vacuum in the tube from the negative to the positive plate.

In Roentgen's day, the nature of cathode rays was still a mystery, though they had been discovered almost forty years earlier. Roentgen had followed the research on cathode rays with great interest for some time, and in October 1895 he had started his own research into the problem at his laboratory at the University of Würzburg in Germany.

It was known that cathode rays produced a greenish luminescence when they struck the wall of the tube. Roentgen, a physicist with a reputation for meticulous experimental technique and acute observation, intended to study this luminescence in some detail. For this purpose he had in his laboratory, lying on a bench near the Crookes tube, a screen coated with fluorescent material. In preparation for his experiment, he covered the Crookes tube with black cardboard, turned on the tube, then turned off the lights in the laboratory, presumably to be sure the cardboard blocked any stray light. It did, but to his surprise the fluorescent screen on the bench began to glow. He turned off the tube. The luminescence faded. He turned on the tube and it appeared again.

Wilhelm Roentgen.
(Deutsches Museum,
Munich)

Clearly the luminescence was caused by the Crookes tube; by implication it was somehow related to the cathode rays produced by the tube. But the rays producing the luminescence could not be cathode rays. It had been known for some time that cathode rays cannot pass through the glass walls of a Crookes tube. Light rays can pass through glass, but not through black cardboard. These mysterious rays behaved differently from anything Roentgen had ever observed. When he held a book between the tube and fluorescent screen, the rays went through the book! Roentgen had discovered a new type of radiation, which he called x-radiation, to denote its unknown nature.

When asked what he thought when he first noticed the fluorescence on the screen, Roentgen replied, "I did not think; I investigated." This attitude explains why Roentgen is honored as the discoverer of a phenomenon that others had probably observed in the previous three decades of work on cathode rays. A number of scientists had noticed fluorescence and the fogging of photographic plates near Crookes tubes, but, unlike Roentgen, they did not investigate. His discovery may have been serendipitous, but, in Louis Pasteur's words, "Chance favors only the mind that is prepared."

For six intense weeks Roentgen repeated and extended his observations of x-rays. He proved that x-rays were produced when the cathode rays struck the plate or the walls of the tube. The energy of motion of the cathode rays or electrons is transformed into x-rays, which spread out in all directions from the place of origin. He demonstrated that, unlike cathode rays, x-rays are not deflected by a magnet. By passing x-rays through different types of material, he showed that they do not reflect or refract from matter nearly as strongly as light rays do.

The most fascinating property of x-rays is their ability to penetrate ordinary matter. They pass through materials that are opaque to visible or ultraviolet light, and they travel about two meters in air before being absorbed. Roentgen showed that this ability of x-rays to pass through an object depends on the density of the object. For example, they pass through wood easily but can be stopped by a small thickness of lead. They pass readily through flesh but not through bones. Roentgen made x-ray photographs of balance weights in a closed box, of the chamber of a shotgun, and of the bones in his hand. Late in December 1895 he took his wife, Charlotte, to his laboratory and made a photograph of her hand. The bone structure and her wedding ring showed clearly. A few days

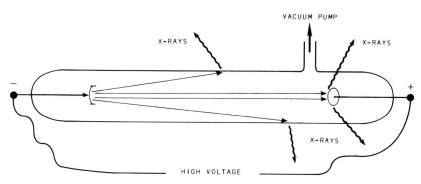

The production of x-rays in a cathode ray tube.

later he communicated the results of his investigations, along with this and other photographs, to the Physical and Medical Society of Würzburg. Within a few weeks, the news had spread around the world that Roentgen had discovered rays that could be used to see through matter.

This discovery not only had profound implications for science but also had clear practical applications for medicine. X-radiation was easy to produce: all one needed was a Crookes tube and a generator to provide the necessary power, both of which were relatively inexpensive and readily available. Medical researchers, amateur scientists, and gadgeteers joined physicists in attempting to understand and apply x-radiation.

Even the guardians of the morals of the masses got involved. To some of them, x-rays represented a moral danger. The lewd and lascivious might use x-rays to look through the clothes people wore! In the New Jersey legislature, it was proposed that the use of x-rays in opera glasses be banned.

In a more practical vein, x-rays were put to use a few months after Roentgen's discovery to produce the first dental radiographs, to locate bullets in wounded soldiers in the Italian-Ethiopian conflict, and to guide the setting of broken bones. The practical application of x-ray technology today has expanded into new areas, such as CAT scanners, airport baggage-inspection machines, and the nondestructive testing of jet engines and other machinery.

Applications to scientific research came more slowly but were equally significant. Since x-rays were not observed to be deflected by a magnetic field, Roentgen assumed, correctly, that they were similar to visible light. If so, they should behave like waves. If they pass through a grating consisting of a number of slits, the original wave should be broken up into a number of smaller waves that will mix or interfere with each other to produce a pattern, called an interference pattern. Roentgen looked in vain for these effects. Eventually he gave up and turned to other areas of research not related to x-radiation. The first Nobel Prize in physics, awarded in 1901, went to Roentgen for his discovery of x-rays. He donated his prize money to support further scientific studies at the University of Würzburg.

In 1912, seventeen years after Roentgen's discovery, Max von Laue, a theoretical physicist at the University of Munich, came up with an idea as to why Roentgen had failed to observe x-ray interference patterns. The wavelength of the x-radiation must be very

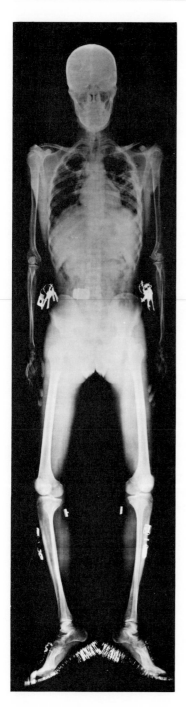

Roentgen's original plate of a man with keys in his pockets. (Deutsches Museum, Munich)

small, so that the waves pass through the slits in the grating without interference. Von Laue reasoned that the regular arrangement of atoms in a crystal could be used as a type of grating for x-radiation, with the distances between atoms assuming the role of the slits. His assistants, Walter Friedrich and Paul Knipping, set up the experiment. They placed a crystal of copper sulphate between a beam of x-rays and a photographic plate, and kept the x-ray beam running for several hours. When they developed the plate, they had a historic photograph. It showed a symmetrical pattern of dark spots, each spot corresponding to the symmetrical arrangement of atoms in the crystal. In one stroke, they had given the first proof that atoms are real and the first proof that x-rays are, like visible light, a form of electromagnetic waves.

Almost immediately, a father-and-son team of English scientists, William Henry and William Lawrence Bragg, showed how x-ray interference patterns, or diffraction patterns as they are also called, can be used to determine the distance between atoms in a crystal. In the years that followed, x-ray diffraction became an indispensable tool for studying the structure of inorganic and organic compounds. In 1946 x-ray diffraction was used to determine the molecular structure of penicillin, thus permitting the synthesis of the antibiotic. Seven years later came the most famous application of x-ray diffraction techniques: the determination of the double helix structure of deoxyribonucleic acid (DNA) by James Watson and Francis Crick.

In the meantime, x-ray research played a central role in fundamental physics. The discovery of x-rays was the beginning of what has come to be called the second scientific revolution. The first revolution was the elucidation of the laws of gravity and motion by Galileo, Kepler, and Newton. The second revolution involved the development of the quantum theory of matter and the understanding of the structure of the atom. X-rays, with their ability to penetrate matter and probe its structure, greatly aided research into the nature of matter. From 1901 to 1917, thirteen Nobel Prizes in physics were awarded for experimental discoveries, four of them for work in which x-radiation played a crucial role. Arthur Compton, a Nobel laureate for research involving x-rays, expressed a common feeling among physicists when he said, "Perhaps no single field of investigation has contributed more to our knowledge of atomic structure than x-rays."

Why were x-rays so vital to this research? Essentially because the energy of x-ray photons is just right. This can be understood roughly in terms of the relationship between the energy and the wavelength of photons. X-rays, which have far more energy than visible-light photons, have much shorter wavelengths — about a thousand times shorter. The wavelengths of x-rays are comparable to the size of atoms. It is a general rule that if you want to study a particular phenomenon, you should choose a tool of a size comparable to the size of the phenomenon. You would not use a steamshovel to investigate an archaeological site, or a toothpick to search for a vein of gold in a mountain. Because x-rays have wavelengths comparable to the sizes of atoms, they are perfectly suited as fine-scale probes of the structure and arrangement of atoms.

The same basic physical reasons that make x-rays so useful for studying atoms also dictated that x-ray astronomy would be a late-developing science. The atmosphere is made of atoms, and x-rays interact with these atoms, so that any x-rays that encounter the

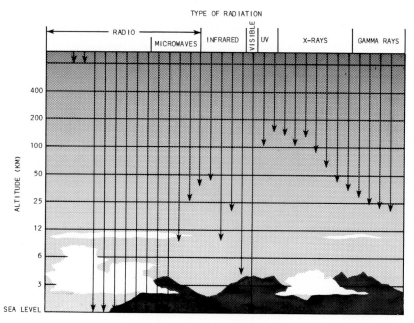

The absorption of radiation by the atmosphere. (From D. Goldsmith, *The Evolving Universe* (Menlo Park, Calif.: Benjamin Cummings, 1981))

atmosphere from space are absorbed high above the earth. This may seem strange, since x-rays pass right through our flesh, for example, and the atmosphere is much less dense than flesh. The explanation is that, even though the atoms in the atmosphere are widely spaced, the total thickness of the atmosphere is so great that all the x-rays coming from outside the atmosphere encounter some of those atoms and are absorbed high above the surface of the earth. (The lower-energy visible-light photons have wavelengths much larger than atoms, so they interact weakly with them and pass unhindered through the atmosphere.) The atmospheric blanket protects the earth from the high-energy radiation from the sun and makes life as we know it possible on this planet. It also makes life difficult for the x-ray astronomer, though, because the only way to observe x-rays from the sun or from the rest of the universe is to place a detector high above the earth on board a rocket or a satellite. The development of the field of x-ray astronomy therefore had to await the dawning of the space age.

4

Pioneers

The pioneering of x-ray astronomy was similar in many ways to any frontier activity. It is on the frontiers that you find the exhilarating mixture of adventure, enterprise, ingenuity, and try-what-works pragmatism. The sponsorship of pioneering activities can usually be traced back to some purely practical concern, such as the opening up of new trade routes. In this, x-ray astronomy was no exception. Its origins can be traced to a suggestion by Thomas Edison that the U.S. government set up a laboratory for the study of advanced communication. The result was the establishment in 1924 of the Naval Research Laboratory (NRL) along the Potomac River in Washington, D.C.

One of the primary missions of NRL was to study the propagation of radio waves and the conditions under which this propagation could be disrupted. Radio waves propagate around the curve of the earth by bouncing off the ionosphere, a layer of charged particles in the upper atmosphere. Edward O. Hulburt directed an ionospheric research program at NRL starting in 1924. Over the course of the next decade he and others showed that radio waves propagate differently at different times of day, different days of the year; furthermore, they noted that the propagation is disrupted during solar flares. They concluded that the sun is the dominant factor in determining the nature of the propagation of radio waves and, by implication, the dominant factor in determining the nature of the ionosphere.

Solar influence over the ionosphere could occur only if the sun produced an invisible flux of high-energy radiation. The visible solar photons do not have enough energy to ionize, that is, to tear apart the electrons from the atoms to form the charged layer. Independent evidence suggested that the sun was a source of high-energy photons. During solar eclipses two outer layers of the sun become visible. The first of these, the one closest to the visible surface of the sun, is called the chromosphere (color sphere); the outer layer is called the corona (crown). An analysis of the visible light from the corona indicates that the corona is a gas with a low density and a very high temperature — about a million degrees Celsius. If so, then the corona should be a strong source of ultraviolet and x-radiation. It is this radiation, suggested Hulburt (and, independently, the Norwegian scientist Lars Vegard), that ionizes the upper layers of the earth's atmosphere to produce the ionosphere.

When this idea was put forth in 1938, it was impossible to test it directly. The best high-altitude balloons could ascend only about 40 kilometers above the surface of the earth, whereas the ionosphere was thought to extend from 50 up to more than 160 kilometers, with most of the x-radiation being absorbed above 100 kilometers. What was needed was a rocket that could carry detectors high above the surface of the earth. Thus, as early as 1938, the scientific need to conduct observations in space was recognized, but the technology for doing this did not exist. At least not on this side of the Atlantic. In Germany, a rocket-development program had been under way since 1930. This program culminated in 1944 with the V-2, a self-propelled rocket perfected by Werner von Braun and used by the German forces in 1944. It was with rockets of this type, captured intact in Germany and operated by the same team that had conceived and built them, that the United States first entered the space age. The U.S. Army shipped 300 boxcars full of V-2 parts back to the United States and announced that V-2 rockets would be available for scientific research.

At NRL, Hulburt and his co-workers immediately began making plans to search for ultraviolet radiation from the sun. On June 28, 1946, they launched their first experiment, an ultraviolet spectrograph located in the nose cone of a V-2 rocket. Herbert Friedman, a member of Hulburt's team, later recalled the result of that experiment: "The rocket returned to earth, nose down, in streamlined flight and buried itself in an enormous crater some 80 feet in diameter and 30 feet deep. Several weeks of digging recovered just a small heap of unidentifiable debris; it was as if the rocket has vaporized on impact."

After that, the scientific payloads were mounted in the tail sec-

A V-2 rocket being readied for launch at White Sands Missile Range.
(R. Tousey)

tion of the rocket, which was severed from the rocket before reentry. With this procedure, the payload was recovered in excellent condition. On October 10, 1946, the NRL group obtained the first ultraviolet spectra of the sun, marking the beginning of space astronomy. While the exploration of the ultraviolet radiation from the sun continued, Friedman and his group, which included Edward T. Byram and Talbot Chubb, pushed on to higher energies.

Their approach was to use selective gas ionization thresholds and filter materials to make Geiger counters that were sensitive to a narrow range of wavelengths. The basic idea is this: a gas, such as argon mixed with carbon dioxide or methane, is put into a box containing one or more high-voltage wires, which are called electrodes. The box is then sealed with a thin "window" made of foil. This window will block out light, but will allow certain wavelengths of ultraviolet or x-radiation to pass through. Adjusting the thickness of the foil windows makes it possible to filter out different wavelengths. Thus, for example, a thin window may admit all wavelengths, from the extreme ultraviolet through hard x-rays, whereas a slightly thicker window will block out the extreme ultraviolet, an even thicker window will also block out the soft x-rays,

Recovering the remains of a V-2 rocket. (R. Tousey)

and so on. Once a high-energy photon penetrates the window, it is absorbed by the gas in the counter. In the absorption process, one of the atoms of the gas ejects an electron. This free electron plows through the gas and liberates other electrons from their atoms. These electrons then drift toward the high-voltage wire, picking up energy and ionizing still more atoms. By the time the electron cloud reaches the high-voltage wire, thousands of electrons have been liberated, and a large electrical signal has been produced. When the detector is operated at very high voltages a saturated signal is produced; that is, the signal is independent of the incoming photon energy. When the detector is operated at lower voltages, the average number of free electrons and hence the strength of the electrical signal is proportional to the energy of the incoming photon. Hence, detectors of this type that are operated at low voltage are usually called proportional counters.

In September 1949 Friedman's group flew a V-2 rocket carrying a

Herbert Friedman.
(Naval Research
Laboratory)

collection of detectors sensitive to extreme ultraviolet and x-radiation. As the rocket rose to 75 kilometers above the earth, the ultraviolet counters registered a strong signal. Above 85 kilometers, the x-ray counters began to register a strong signal. Visible-light photocells confirmed that the counters were pointing toward the sun when they registered x-rays. Slightly more than a decade after the predictions of Hulburt and Vegard, Friedman's group had proved that the sun was a source of x-rays and that these x-rays were absorbed 90 or more kilometers above the earth, producing the upper ionospheric layers.

After this historic flight, the group attempted much more ambitious experiments, but things did not go so smoothly. In Friedman's words, "Our beginner's luck deserted us. A 1950 V-2 exploded on the pad before our eyes, the biggest bonfire we had ever seen." Shortly thereafter, they launched another payload, this time on a Viking rocket. But the rocket spun too rapidly, and the instrumentation was torn from its mountings.

These and other failures naturally spurred a search for a more reliable rocket. James Van Allen, who would later gain fame as the discoverer of the Van Allen belts of energetic charged particles around the earth, initiated the development of the Aerobee rocket. This was a small rocket, only about 6 meters long and 160 pounds. It was inexpensive, costing only about $30,000, and proved to be extremely reliable. During the period from 1948 to 1976, the Aerobee had a success record of greater than 90 percent.

With the Aerobees, Friedman's group undertook a detailed study, lasting more than a decade, of the x-ray emission from the sun. By the late 1950s they had studied the sun through an entire sunspot cycle of eleven years. It was well known by then that the number of sunspots waxes and wanes with a fairly well defined period of eleven years. As the number of sunspots increases, so does the incidence of solar flares, which come from regions around sunspots called active regions. It had been known for some time that shortwave radio transmission is disrupted during solar flares. Since this transmission involves the ionosphere, which is produced by high-energy radiation from the sun, it seemed probable that solar flares produced x-rays. In 1956 Friedman set out to prove this hypothesis by launching x-ray detectors during a solar flare.

This was not simple. Flares rise to maximum in only a few minutes and may occur only once a day or even less often, at times that cannot be predicted. Therefore the only way to observe them is to

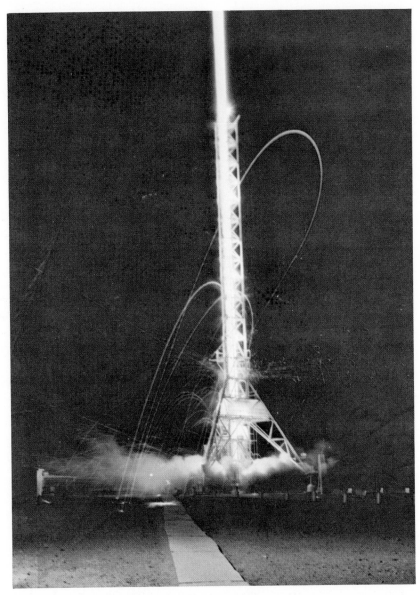

The launch of an Aerobee rocket. (U.S. Navy)

have a telescope or detector available on a standby basis. Aerobee rockets could not be used as standbys. Their liquid propellants could not be stored in a rocket tower for more than a few hours. To catch a solar flare you might have to wait five or ten hours or more. What was needed was a rocket that used solid propellants. These propellants, explosive powders, can be stored for long periods, but they provide a much lower thrust. The only solid-propellant rocket available at the time, the Deacon, could reach an altitude of only 40 kilometers when launched from the earth. If the Deacon could be launched from a high altitude, however, the reduced atmospheric drag would allow it to reach a much greater height.

James Van Allen devised an ingenious way to send rockets higher above the earth: the rockoon, a combination of a rocket and a balloon. They would use a balloon to carry the Deacon almost 100,000 feet above the surface of the earth and launch it from there. With this arrangement, the rocket could climb to well over 100 kilometers. Friedman was able to convince the Navy to use the rockoon to observe solar flares, and his group set out to sea off the coast of southern California with ten rockoons. Early in the morning of each day they would launch a huge polyethylene balloon with a Deacon rocket dangling below. If astronomers at the High-Altitude Observatory in Colorado or the Sacramento Peak Observatory in New Mexico detected the onset of a solar flare, they would immediately notify Friedman's group by teletype, and the rocket would be fired. If no flare occurred during the day, the rocket still had to be fired at the end of the day, to keep it from drifting out of the prescribed firing zone. They had ten rockets, so if every rocket worked they had ten days in which to catch the eruption of a solar flare.

On the first day out a flare occurred, but the rocket failed to ignite. Twice. The problem was quickly traced to the near-vacuum conditions at an altitude of 100,000 feet. There simply was not enough oxygen to make the explosives work. Friedman called on James Kupperian, Jr., to solve the problem. Kupperian was the on-board expert in high-altitude explosives because, according to Friedman, "In his teenage years he had acquired considerable expertise in making gunpowder from the ingredients in his chemical set." After one more failure, Kupperian found a solution. Now all they needed was a solar flare. Five days passed without a flare. Finally, with only one rockoon remaining, a weak flare occurred. The rocket went up, and for the first time, the intense x-ray emission from solar flares was observed.

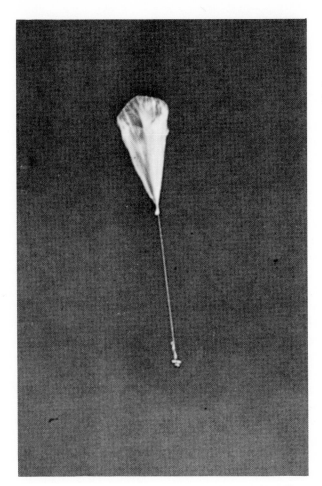

A rockoon.
(R. W. Kreplin)

By 1958 the rockoons were obsolete; a two-stage solid-propellant rocket using a Nike booster had been developed. The excitement of doing rocket astronomy did not diminish. On October 12, 1958, the path of a total eclipse of the sun crossed the South Pacific. Friedman and his group were there, on the tiny coral atoll of Puka Puka in the Danger Islands. Their plan was to launch a series of six rockets as the disk of the moon occulted the sun. They hoped to isolate discrete sources of x-rays associated with sunspots, and to see how far the x-ray-emitting regions extended above the surface of the sun.

The group set up six rockets on the helicopter deck of the U.S.S. *Point Defiance.* The rockets were mounted on tripods welded to plates of quarter-inch steel; these served to deflect the blast, which had the intensity of a thousand-pound bomb. The idea of setting off the rockets aboard a ship was a bit disconcerting to Friedman, who admitted, "we had some nightmares of the good ship *Point Defiance* going to the bottom when we exploded the first Nike."

But the ship withstood the blasts in good order, and they fired the first four rockets without incident. Then the button was pushed to fire the fifth rocket. Nothing happened. Apparently the blasts from the previous four launches had loosened a plug that carried the starting signal to the rockets. The eclipse was progressing and precious seconds were ticking away. Friedman recalled the hectic events that followed: "While we watched with bated breath, Don Brousseau of the NRL group raced up the ladder to the helicopter deck, climbed the tripod frame, and reinserted the loose plug in the rocket. He then dashed for the nearest guntub and huddled behind its armor-plated protection while the rocket was fired, no more than 50 feet away."

The mission was a success. The change of the measured x-ray fluxes in the course of the eclipse showed that the x-ray emission from the sun was concentrated in active regions around sunspots, and that the x-ray-emitting layers extended well above the surface of the sun. By the end of the 1950s, Friedman's group had shown conclusively that the appearance of the sun at x-ray wavelengths was markedly different from its appearance at visible-light wavelengths, and that x-ray observations were essential for understanding such high-energy phenomena as solar flares and the development of the active regions that produce solar flares. Perhaps more important, Friedman's group had demonstrated that doing astronomy from space produced a rich return of scientific results, a return that was well worth the risks involved.

Not everyone agreed. In fact, most astronomers disagreed. Some simply preferred the independent, solitary mode of research that is possible in an optical observatory but not in space astronomy, which requires a large team. Others found the risks too high. Still others ventured into the field briefly but were so disheartened by rocket disasters that they gave up and went back to their earth-bound observatories. Another reason, undoubtedly, was that few astronomers believed there was any prospect in the near future of extending x-ray astronomy beyond the solar system. The x-ray luminosity of the sun is a million times smaller than the optical luminosity. If the sun was any guide, then to detect x-rays from the stars

Six Nike-Asp rockets poised on the deck of the U.S.S. *Point Defiance* to observe a solar eclipse. (U.S. Navy)

would require a detector thousands of times more sensitive than ones already in existence.

For these reasons, Friedman's group was virtually alone in the 1950s as they pioneered the x-ray studies of the sun. But in October 1957 an event had occurred that ensured that they would not be alone in the field for much longer. In that month the space scientists of the Soviet Union launched the first artificial satellite, *Sputnik 1.*

5

The Discovery of an X-Ray Star

In 1955 the top scientist in the Soviet space program, Leonid Sedov, in a rare appearance outside the USSR, attended a congress of the International Astronautical Federation (IAF) in Copenhagen. At the congress he announced the USSR's intention to launch the first in a series of artificial satellites for the purpose of conducting space research. Judging from the reaction just two years later, it seems safe to conclude that virtually no one took him seriously. But on October 4, 1957, during another congress of the IAF, the Russians announced, this time by telegram, that *Sputnik 1*, a satellite weighing 83.6 kilograms, was in orbit. A month later *Sputnik 2*, which weighed more than 500 kilograms and carried a dog, was launched; six months after that came *Sputnik 3*, which weighed 1327 kilograms. There seemed little doubt of the Soviet dominance in space technology.

In the United States, suggestions as to what to do about this unhappy state of affairs ranged from President Eisenhower's reassurances that it was nothing to get upset about, to talk of a preemptive nuclear war. Fortunately, a view between these extremes prevailed, namely that the United States should embark on a crash program of research and development in space science and technology. The next question was who should be in charge of the space program. The Pentagon argued, with the support of large industrial companies, that the military was the natural choice; it had been doing space research since the end of World War II, and security was vital, military spokesmen claimed, in the development of a technology that mainfestly had strategic importance. On the other side, many scientists and engineers, as well as influential members of

Congress, argued that the space program should be controlled by civilians, in an atmosphere where information could flow freely and good ideas would rise to the top. At that time the Pentagon did not enjoy the best of relations with Congress. Sputnik had been a public relations coup for the Soviet Union as well as a technological one, and the prestige of the U.S. military suffered. There was also a feeling in Congress that a "missile mess" was being created by interservice rivalries within the Pentagon. Moreover, President Eisenhower was becoming increasingly distressed by the power and unmanageability of what he later called the military-industrial complex. He was not inclined to assign it a new and major responsibility. These reasons, together with lobbying from both inside and outside the government, led to a decision in favor of a civilian agency.

The decision was that some space projects having specifically to do with national defense, such as missile development and some scientific work, would continue to be funded by the military, but the major U.S. efforts in space research and technology would be directed by the National Aeronautics and Space Administration (NASA). Established by act of Congress and signed into law by President Eisenhower on July 29, 1958, NASA began operations just three days short of a year after the launch of *Sputnik 1.* The subsequent triumphs of the American space program in both science and technology have demonstrated beyond a shadow of a doubt the wisdom of the decision to create a civilian agency. Its accomplishments are especially impressive when compared with those of the USSR's science program, which is dominated by the military. The current trend of closer cooperation between military and civilian space activities, brought about by the substantial use of the Space Shuttle Transportation System by the Department of Defense, is cause for concern. It may be that the great advantage of a separate civilian space program, which NASA has allowed the working scientists to help shape, will be lost in the future.

NASA recruited James Kupperian, Jr., of the NRL x-ray astronomy group to be head of the astronomy branch at the Goddard Space Flight Center in Greenbelt, Maryland. Kupperian brought to the job his interest in x-ray astronomy, and he immediately sought to interest the scientific community in space research. He found little enthusiasm among astronomers, probably because of the reasons discussed earlier, the primary one being that research in space is an inherently risky business with doubtful returns. As a result, only two groups were funded initially to carry out an x-ray program: one at Rochester University under the direction of Malcolm Savedoff, and one at Lockheed Missiles and Space Company under the direction of Philip Fisher.

Meanwhile, the National Academy of Sciences had established the Space Science Board to "survey the scientific problems, opportunities and implications of man's advance into space." The Space Science Board was to consist of eminent scientists who would serve terms of a few years each and would draw upon the advice of numerous subcommittees in advising NASA and formulating a broad policy on space research. From its inception the Board has played a key role in influencing the direction of research. Its early interest in surveying the sky at wavelengths shorter than the visible helped to pave the way for the development of x-ray astronomy. In October 1958 John Simpson, chairman of the Committee on Physics of Fields and Particles in Space, suggested x-ray and gamma-ray mapping of the sky. Laurence Aller, a member of the Committee on Optical and Radio Astronomy, pointed out in 1959 that, whereas interstellar hydrogen would absorb most of the ultraviolet radiation from stars, x-rays could penetrate distances of thousands of light-years. Leo Goldberg, a member of the same committee, advocated the development of x-ray instrumentation in view of its potential benefits to astronomy.

These opinions helped to establish a friendly environment for the development of x-ray astronomy, but the views of another member of the Space Science Board were to have a more immediate and far-reaching impact. Bruno Rossi was a member of the Committee on Space Projects. Rossi, one of the most distinguished experimental physicists of the day, had done pioneering work in cosmic ray physics, worked on the Manhattan Project, and written classic books on cosmic rays and optics. In addition, he had, in his own words, "a deep-seated faith in the boundless resourcefulness of nature, which so often leaves the most daring imagination of man far behind." This credo motivated him to consider the possibility that a program in x-ray astronomy might be worth instituting immediately instead of at some time in the future. Rossi provided the initial suggestion to start an x-ray astronomy program at American Science and Engineering, a private company in Cambridge, Massachusetts, near MIT. Also, through his presence in Cambridge and

Bruno Rossi.

his continued enthusiasm for x-ray astronomy, he contributed greatly to the climate of intellectual fervor and discussion in the exciting years ahead.

American Science and Engineering (AS&E) was formed in 1958 by a small group that included two of Rossi's former students at MIT: Martin Annis, who became president, and George Clark. Rossi soon joined the enterprise as chairman of the board and scientific consultant. Clark, like Rossi a professor in the physics department at MIT, served on the board of directors and as a scientific consultant. One of the company's goals was to do scientific research and development on contracts from government agencies. The Department of Defense was one of the primary sources of these contracts, and AS&E was involved in studying effects of nuclear weapons when Riccardo Giacconi joined the company in early 1959 to start a program of space research.

The 28-year-old Giacconi had done his thesis on elementary particles at the University of Milan, working with a great scientist and teacher, Giuseppe Occhialini. After finishing the thesis, he taught at the University of Milan, where he designed and built the largest multiplate cloud chamber in Italy. Then, at Occhialini's suggestion, he came to the United States on a Fulbright grant to work with Robert Thompson at Indiana University. Later, at Princeton, he led experimental work in detection of Cerenkov radiation from μ mesons in air, participated in the development of scintillation chambers using newly invented image intensifiers, and conducted (with Herbert Gursky and Fred Hendel) an experiment to search for new mesons. Like many other cosmic ray physicists at the time, he found the potential of cosmic ray research to be dwindling with the advent of powerful manmade particle smashers. The large fluxes of particles available from these machines on command made the painstaking collection of a few rare events, which was the norm in cosmic ray research, appear futile and quaint. Giacconi had spent two years collecting eighty proton interactions for his thesis work at the Testa Grigia Laboratory, an aerie located 11,000 feet above sea level at the foot of the Matterhorn. He had dreamed of magnetic funnels to increase the flux of particles on his instrument, but such schemes had been only wishful thinking. In 1959 he was ready to redirect his research interests.

At AS&E, Giacconi met Clark, who suggested several lines of space science research having to do with measuring the properties of charged particles in the Van Allen radiation belts. Giacconi began work on some of these experiments and also, in aid to AS&E's existing program, began studying ways to build a directional detector to measure gamma rays emitted in a nuclear bomb explosion in space. Then in September 1959 he attended a party at Rossi's house and met Rossi for the first time. Rossi suggested that a venture into x-ray astronomy might prove very fruitful, not because of any theoretical predictions, but because nothing was known and there was the possibility for major new discoveries.

Impressed by Rossi's suggestion, Giacconi almost immediately began an x-ray astronomy program at AS&E. He made a thorough survey of the meager literature. Giacconi and Clark, with some assistance from Stan Olbert at MIT, made theoretical estimates of possible sources of x-rays outside the solar system. They considered solar-type stars, flare stars, stars with magnetic fields ten thousand times larger than that of the sun, and the Crab Nebula, the gaseous remnant of a supernova explosion, which was known to be a strong source of radio waves. The estimates were necessarily very uncertain, but it seemed highly unlikely that the existing detectors would pick up any x-rays from these sources. If the AS&E team did find sources, then to study these sources in detail they would need x-ray

Riccardo Giacconi, Herbert Gursky, and Fred Hendel in the Palmer Physics Laboratory, Princeton University.

detectors hundreds or thousands of times better than any that had been built before. At first Giacconi felt discouraged.

What was ultimately needed, Giacconi concluded after considering the problem in more detail, was a means of focusing the x-rays from a large collector onto a small detector. In other words, an x-ray telescope. Because of their short wavelength and high energy, x-rays encountering a mirror at a large angle do not reflect, but penetrate the surface and are absorbed, in much the same way a stream of bullets would slam into a wooden wall. But just as bullets hitting the wall at a grazing angle will ricochet so too will x-rays striking polished glass mirrors at grazing angles be reflected with high efficiency. Giacconi found a description of this effect in the *Encyclopedia of Physics* by Siegfried Flugge, and he immediately realized that a parabolic mirror could be an efficient x-ray collector. It could be done! The small numbers of x-rays expected from x-ray sources would not make them impossible to study.

A search of the literature revealed that Hans Wolter, a German physicist, had studied the problem of focusing x-rays about a decade earlier in an attempt to develop an x-ray microscope. He had showed that x-ray images could be formed by causing x-rays to be reflected at grazing angles from properly shaped surfaces. Wolter's proposals had no practical application for microscopy; it was simply too difficult to construct optical surfaces of the required precision on the small scale required. For astronomical applications, however, large mirrors are needed, and the fabrication of the mirrors is no more difficult than for any other astronomical telescope.

Giacconi discussed the idea of an x-ray telescope with Martin Annis, who suggested calling in Bruno Rossi. Rossi embraced the concept enthusiastically and pointed out that the effective area of the telescope could be increased by nesting several mirrors in concentric cylinders. In 1960 Giacconi and Rossi published a description of the technique. Giacconi immediately set to work to develop an x-ray telescope and submitted a proposal to NASA for experimental investigations into the fabrication and testing of an x-ray telescope.

Meanwhile, realizing that the construction and flight of an x-ray telescope were some years in the future, Giacconi sought a more immediate solution to the detector problem. The approach was to do the best that could be done with instruments similar to, though more sensitive than, those already being used by the NRL group. He proposed to NASA an experiment to look for x-rays from stars. NASA was not interested.

Because of the several classified programs that AS&E was conducting, it seemed natural to turn for support to the Air Force Cambridge Research Laboratories (AFCRL), located some twenty miles west of Boston at Hanscom Field. It had occurred to the AS&E group that the moon might be a stronger source of x-rays than the

Experimental apparatus for testing the first model of an x-ray telescope. The glass cylinder has the inside surface polished as a cone. (American Science and Engineering)

anticipated weak stellar emission; fast particles streaming from the sun might produce x-rays when they struck the surface of the moon. If so, then x-ray observations of the moon would be a way to monitor the intensity of particle streams from the sun. AFCRL agreed to fund a program of lunar and solar x-ray studies. The AS&E group started work in early 1960 in a building that had been a garage for milk trucks before it was converted into a laboratory. Giacconi was able to hire an electronic and mechanical technician and to have two more physicists, Frank Paolini and Norman Harmon, assigned to the project. The group was strengthened a year later by the addition of Herbert Gursky.

The researchers quickly realized that one of the basic problems would be the background noise produced by cosmic rays, energetic charged particles that continually bombard the top of the atmosphere. These particles would enter the counter and produce signals similar to those produced by x-rays when the counter was used in the saturated (high voltage) mode. The background caused by cosmic rays is about one count per square centimeter of detector area. This is millions of times less than the number of counts produced by x-rays from the sun, so the cosmic ray background noise poses no problem for observing solar x-rays. But for observing sources that were expected to produce about one photon per square centimeter of the detector, it was clearly a serious problem. A method was needed to tell when a cosmic ray hit the counter and when an x-ray hit it.

The key lies in the much greater energy of the cosmic rays. Because of their high energy, the cosmic rays will go all the way through the detector, leaving behind a trail of ionized gas that produces a signal that is essentially indistinguishable from the signal produced by an x-ray. The difference is that an x-ray will be absorbed before leaving the detector. Therefore, placing additional cosmic ray detectors around the x-ray detector makes it possible to distinguish between the x-rays and the cosmic rays. A cosmic ray will produce a signal in both the x-ray detector and the surrounding cosmic ray detectors, whereas an x-ray will produce a signal only in the x-ray detector. Discarding signals produced coincidentally in the two types of detectors reduces the cosmic ray background to a tenth of its previous level. This technique is called anticoincidence.

An obvious way to increase the strength of the signal is to increase the area of the detector. Proportional counters are widely used in x-ray technology and its applications, but a search for larger counters turned up none with a sensitive area larger than about one square centimeter, roughly the size of a penny. The AS&E group decided to use one of these for an experiment to be flown on a small rocket and meanwhile to develop larger counters in their laboratories.

In June 1960 Giacconi, Paolini, and two AS&E technicians, Al deCaprio and Tom Quinn, traveled to Egland Air Force Base in Florida with their first x-ray astronomy payload. The experiment failed when the Nike-Asp rocket engine misfired. A second proposal was submitted, and another flight was scheduled for October 1961. By then Paolini had developed proportional counters with an effective area ten times larger than any previously available. These counters had x-ray windows made of sheets of mica only a few thousandths of a centimeter thick. This technical improvement, together with the use of anticoincidence and single-photon counting, gave the AS&E group an instrument a hundred times more

OBSERVATIONAL TECHNIQUES

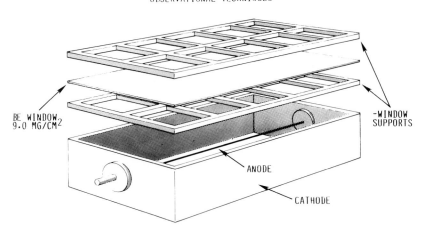

Schematic layout of a thin-window gas proportional counter. X-rays enter the detector through the beryllium windows and interact with the gas inside the counter. The electrons produced by this interaction are accelerated toward the anode. As they move toward the anode, they interact with the gas to produce more electrons. All the electrons are collected at the anode and produce an electronic signal whose strength is proportional to the energy of the incident x-ray.

sensitive than its predecessors. Also, the large field of view (120 degrees) used in the experiment greatly increased the probability of observing a celestial source. The U.S. Air Force, through AFCRL, made available a much larger rocket, one of the dependable Aerobees, and three of the larger counters were included in the experiment payload. The AS&E group journeyed to White Sands Missile Range in New Mexico, where they carefully bolted the payload into the body of the rocket, made last-minute checks, closed the doors that would be jettisoned when the rocket was safely in the upper atmosphere, and waited.

On October 21, 1961, the Aerobee rocket lifted off the desert floor and climbed steadily to an altitude of 200 kilometers. As Giacconi, Gursky, and Paolini watched the strip chart record the incoming data, their hearts sank. No data were coming in. The doors to the rocket had stuck and failed to open. The experiment was a failure. They scheduled another launch for eight months later, in June.

The inner portion of the counter flown in June 1962. (American Science and Engineering)

Meanwhile, President Kennedy, in response to nuclear weapons testing by the Soviet Union, decided in the fall of 1961 to inaugurate nuclear weapons testing in the United States. In October 1961, AFCRL asked AS&E if they wished to undertake a crash program to measure the effects of bursts of nuclear weapons at high altitudes. The program was to be completed by March 1962. Money was no object, but time was precious. AS&E accepted the project and began to prepare payloads to measure electrons, x-rays, and gamma rays produced by nuclear weapons explosions in the atmosphere. From the fall of 1961 to the summer of 1962, Giacconi's group expanded from half a dozen to seventy or eighty people. They designed, built, tested, integrated into vehicles, and launched twenty-four rocket payloads in an eight-month period. They built and integrated six satellite payloads. In about 95 percent of the cases the payloads worked properly and the experiments were successful. This experience molded the AS&E group into a loyal, dedicated, and highly skilled team. They had acquired a reputation and a confident, "can do" attitude; they were also aggressive in seeking funds to expand their research program. By the summer of 1962 they were ready to try again with the x-ray astronomy experiment.

One minute before midnight on June 18, 1962, a rocket carrying three proportional counters, each with an effective area about the size of a credit card, was launched from White Sands. The moon was one day past full, in the southeast, about 35 degrees above the horizon. The rocket reached a maximum altitude of 225 kilometers and was above 80 kilometers for a total of 5 minutes and 50 seconds, traveling almost straight up, with a slight, 3-degree tilt toward the north. It was spinning rapidly along its long axis, making 2 revolutions per second. Two of the three counters operated properly. The third counter was apparently undergoing electrical arc discharges, and had to be disregarded.

As the doors to the rocket opened, the counting rate in the two working counters increased rapidly as the rocket moved into the upper atmosphere. A large peak in the rate was observed every time the counters spun past a certain point in the southern sky. Since the signals were continuously recorded on a strip chart, it was immediately obvious that there was a strong source in that part of the sky. The jubilation in the blockhouse was tempered, ironically, by the strength of the source. It was much stronger than anyone had anticipated any possible extrasolar x-ray source would be. Had something gone wrong?

The experiment payload flown by the AS&E-MIT group in June 1962, with which the first cosmic x-ray sources were discovered. (American Science and Engineering)

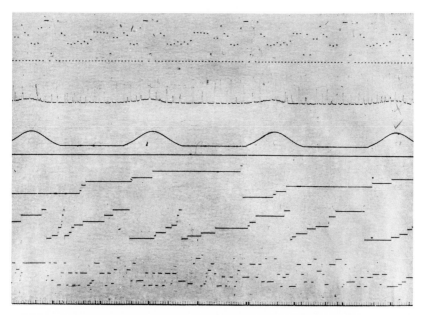

A portion of the telemetry strip chart obtained during the June 1962 flight. (R. Giacconi et al., *Physical Review Letters 9* (1962), 439)

Gursky's reaction was typical of the scientists present: "I saw the peak; the senior technician who was standing there with the data saw it. He said: 'Ah, there it is!' Now I didn't feel that way about it. I knew what the rates should have been and I knew we would have to add all the data together before we had a chance to determine the signal accurately. So I felt we were in trouble, but I didn't know why."

The group returned to Cambridge and began to analyze the data, still uncertain of what they had seen. One concern was that they had detected the moon, but not in x-rays. Perhaps, they worried, the carbon black coating used to shield the counters from ultraviolet radiation had chipped off, and ultraviolet radiation from the moon had leaked through to produce the signal. But the signal was not centered on the moon; it came from a position 30 degrees away from the moon. There was no signal from the moon, or from another known strong ultraviolet source in the constellation of Virgo. Ultraviolet contamination could be ruled out.

The location of the source near the direction of the south magnetic pole was also cause for worry. Could the signal have been produced by charged particles in the upper atmosphere, spiraling along magnetic-field lines? Little was known at that time about the numbers and properties of charged particles in the upper atmosphere, so this was a major concern. Three factors, however, gave the group confidence that they had indeed detected an extrasolar x-ray source. The first was that the signal was sharply peaked. A sharply focused beam of charged particles would be required to mimic what was observed; that such a beam should stream out of the radiation belts and encounter their detectors seemed improbable, since the particles in the radiation belts were thought to be moving in a very broad beam. Secondly, about 60 degrees to the east of the peak, in the direction of the constellation of Cygnus, was another, smaller peak. This secondary peak could not be explained by particles spiraling along a magnetic field. Particles spiraling along a magnetic field would surely show symmetry with respect to the magnetic field, in contrast to what was observed.

The third factor illustrates how, in science as in life in general, the right thing is sometimes done for the wrong reason. While the AS&E group was analyzing the x-ray data, a group of scientists from MIT had just returned from studying high-energy gamma rays in Bolivia. This team, which included George Clark, William Krau-

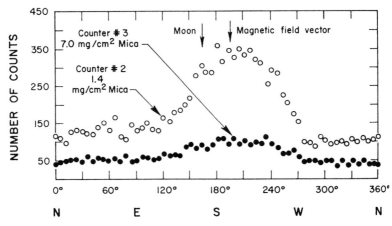

Azimuthal distributions of recorded counts from Geiger counters flown during June, 1962. (R. Giacconi et al., *Physical Review Letters 9* (1962), 439)

shaar, and Minoru Oda, met with the AS&E group for a joint discussion of results. Oda, a Japanese physicist, noticed that the x-ray peak seemed to coincide with a small peak in the gamma-ray data. This coincidence gave the AS&E group comfort. At least there was some celestial feature that seemed to correlate with the x-ray results. Ironically, several months later the gamma ray peak was shown to be spurious. In the meantime, though, it buoyed the confidence of the x-ray group. By late August, two months after the discovery flight, Giacconi, Gursky, Paolini, and Rossi were sufficiently confident of their conclusions to present the results at the Third International Symposium on X-Ray Analysis, held at Stanford University. Herb Friedman was in the audience and heard Giacconi describe the discovery of the first source of x-rays outside the solar system, a source that was just below the level of detectability of a search made by Friedman's NRL group a few years earlier as a by-product of their studies of x-rays from the sun. Clearly they had just missed making this discovery. After the presentation, Friedman introduced himself to Giacconi and congratulated him.

Shortly after the Stanford symposium, the news of the discovery of the first cosmic source of x-rays was published in a letter to the editor of the *Physical Review*. The scientific community's immediate reaction to the announcement was cautious. Kenneth Pounds, who would soon enter the field of x-ray astronomy, remembers: "There was a certain amount of disbelief—or at least we wanted confirmation. I think everybody did."

The confirmation came quickly. The AS&E group verified their results in rocket flights in October 1962 and June 1963. In April 1963, Friedman's group provided independent verification and detected a strong source in the direction of the Crab Nebula. There could now be no doubt: a new field of astronomy had been born.

6

The Riddle of the X-Ray Stars

The first cosmic source of x-rays to be discovered was given the prosaic but practical name Sco X-1, meaning the strongest x-ray source in the constellation of Scorpius. Of course it was much more than that. It was the nugget that proved that a rich vein was waiting to be mined. Other strong sources were soon detected, so suddenly x-ray astronomers had a set of objects they could study with existing technology. They did not have to make their instruments a million times more sensitive, as they had previously thought. Although x-ray astronomy remained a risky venture, with the uncertainties of a developing space technology, the rewards were no longer uncertain. Now they were real and immediate. Other groups besides AS&E rushed to get into the field.

The first of these was Herb Friedman's group at the Naval Research Laboratory. This team must have been bitterly disappointed to have lost the opportunity to discover the first x-ray star. Their program of observing solar x-rays had brought them to the verge of such a discovery several years before; they had even looked briefly, but had seen nothing definite. Although they had fully intended to repeat these efforts with more sensitive instruments, it had not seemed to be a high-priority item, and they had concentrated their efforts in other areas that appeared more fruitful. Undeterred by this disappointment, now they quickly joined the hunt. They developed and flew in eight months a detector that was ten times more sensitive than the original AS&E detector. They verified the existence of a strong source in Scorpius, and they detected a new source in the Crab Nebula, the remnant of an exploded star.

The two groups originally funded by NASA to do x-ray astronomy did not do as well as the NRL team. Malcolm Savedoff's group at Rochester University abandoned x-ray astronomy after encountering technical difficulties in their attempt to build an instrument that could withstand the intense vibrations of launch and the high vacuum of space. Philip Fisher's team at Lockheed also ran into difficulties. Subcontracting delays and other problems forced them to postpone their first scheduled launch from December 1961, six months before the successful AS&E flight, to April 1962; it was September 1962 before they finally got off the ground. Even then, the experiment was plagued by spurious high count rates that invalidated the scientific results. The cause of the high count rates was never fully understood; it could have been a recent high-altitude nuclear explosion, which greatly increased the number of high-energy charged particles in the upper atmosphere, or it could have been a problem with the detector. In any event, the high rates served as a reminder of the difficulties and hazards inherent in doing x-ray astronomy.

The Lockheed group recovered from their slow start and made important contributions to x-ray astronomy. The most significant of these was the introduction of a new technique for scanning the sky. The earliest experiments were performed with spin-stabilized rockets: rockets that spun rapidly about their long axis. Left alone, such a rocket will maintain a fixed heading; the detectors, which are looking out the side of the rocket, will sweep in a great circle as the rocket spins. On an October 1964 flight, Fisher's group became the first to use a rocket whose pointing was controlled by gas jets; with this system, the rocket could be made to perform a slow, highly sensitive scan of a small region of the sky. This technique, which soon became standard practice among all groups, made it possible to locate sources of x-ray emission with much greater precision.

By 1967 there were a dozen or more groups doing x-ray astronomy. Through the combined efforts of the groups at NRL, Lockheed, AS&E, and MIT, a group at the Lawrence Livermore Laboratory led by Frederick Seward, and a University of Leicester group led by Kenneth Pounds, more than thirty sources had been found. Most of these sources appeared to be associated with some bizarre, x-ray-emitting star whose exact nature was unknown. About half a dozen sources could be identified with clouds of gas and high-energy particles produced by supernova explosions; among these so-called supernova remnants was the Crab Nebula. Some sources had been observed to flare up and disappear entirely. One or maybe two or three were associated with galaxies millions of light-years away.

The first indication of the potential richness of this new field came, not from a study of Sco X-1 or one of the other strong sources, but from a study of the background radiation that was observed when the detectors were looking away from the sources. In the flight of June 1962, when Sco X-1 was discovered, evidence for a uniform background glow of x-radiation was also discovered. This background is the reference level against which x-radiation from stars and galaxies must be observed. If we look at the night sky at visible-light wavelengths, we see many bright spots, the stars, plus a few diffuse patches of light (produced by galaxies or clouds of glowing gas) on a black background. At x-ray wavelengths, the equivalent of the black night sky does not exist. We see instead a bright, diffuse glow that is the same in all directions. The uniformity of the background x-radiation means that it must originate far outside our galaxy. This discovery has profound implications. It means that studies of the x-ray background can help us to answer cosmological questions such as whether the universe is finite or infinite.

Fred Hoyle, an inventive British astrophysicist who has left his mark on just about every field of astronomy, was quick to appreciate the significance of the x-ray background radiation. Along with Herman Bondi and Thomas Gold, Hoyle had developed a steady-state model for the universe. According to this theory, the universe, as viewed on a scale of billions of light-years and over billions of years, is essentially the same everywhere and always. The observed expansion of the universe appears to contradict this assertion: the galaxies in the universe are observed to be rushing away from one another; therefore they are getting farther apart all the time and space must be getting emptier all the time. The steady-state theory avoided this difficulty by postulating that matter is created continuously in intergalactic space, so that the density of matter in space remains the same in spite of the cosmological expansion. Because of the elegant concept of a universe that was uniform in both space and time, and because of the eloquent advocacy of Bondi, Gold, and Hoyle, the steady-state theory enjoyed wide popularity in the fifties and early sixties.

Hoyle claimed that the steady-state universe predicted an x-ray background that was in rough agreement with the observations. According to the steady-state model, matter is created in the form of neutrons, unstable particles that decay in a matter of minutes to protons and energetic electrons. The energy of these electrons goes into producing a hot intergalactic gas. The radiation from this gas is primarily in the form of x-rays, hence the x-ray background radiation. This was a compelling argument, but it was wrong. When Robert Gould and Geoffrey Burbidge of the University of California at San Diego made a more detailed calculation of the x-radiation expected from the hypothetical hot intergalactic gas, they found that Hoyle's estimate was off by a large factor. The x-ray background expected from the steady-state universe was a factor of seventy *larger* than the actual x-ray background observed. This was, in the words of Philip Morrison of MIT, "the beginning of the rout of the steady-state theory." Support for the steady-state theory in the scientific community declined steadily, and in 1968, when Arno Penzias and Robert Wilson discovered microwave background radiation, it collapsed.

X-ray astronomers have studied the x-ray background radiation for more than two decades and have discovered sources that account for much of it. Nevertheless, a detailed understanding of this radiation remains just beyond the grasp of astronomers, though we are clearly getting closer to the solution (see Chapter 13). In the meantime, the question of the nature of Sco X-1 and the other sources of cosmic x-rays was attacked on two fronts. The sky was surveyed with x-ray detectors to see what was there, and meanwhile certain individual sources were studied in detail.

Both the surveys and the individual studies indicated that x-ray stars were very strange objects indeed. They were apparently giving off a thousand times more energy in x-rays than in visible light. This is just the reverse of the situation on the sun, where the visible luminosity is a million times more intense than the x-radiation. In general, x-rays are produced by high-energy electrons, either in a beam or (as in the sun's corona) in a very hot gas, with temperatures measured in millions of degrees. One possibility suggested itself. Most stars have surface temperatures of thousands of degrees: the temperature of the visible surface of the sun is six thousand degrees Kelvin; red stars such as Betelgeuse have surface temperatures of about four thousand degrees; and blue stars such as Rigel have temperatures of about ten thousand degrees. But what if x-ray stars had surface temperatures of millions of degrees?

A star's luminosity depends on both its temperature and its surface area. The dependence on the temperature is strong. A star that had the same surface area as the sun but was twice as hot would radiate sixteen times as much energy. Therefore a star with a surface temperature of about ten million degrees would have an enormous luminosity unless it had very small surface area. The implied diameter of the x-ray star was about six miles.

Could stars only six miles in diameter exist? If so, could they have surface temperatures of ten million degrees or more? In the early sixties, the answers to these questions were not known, but there was a body of theoretical work that suggested that both answers might be yes.

The surface of a star is held up by a balance of internal pressure against gravity. In the normal course of a star's evolution, the internal pressure is provided by energy from nuclear reactions deep in the core of the star. When nuclear reactions stop producing energy, the pressure drops and the star falls in on itself. A star about the size of the sun will collapse into a ball about the size of the earth, or about one one-hundredth the original diameter of the star. These collapsed stars, which were discovered early in this century, are called white dwarf stars, because they are so small and the heat generated by the collapse has made them white hot. A sphere of white dwarf material the size of the period at the end of this sentence would weigh about two pounds — about a hundred thousand times more than a lead sphere of the same size.

In the 1930s the Cal Tech astronomers Fritz Zwicky and Walter Baade conjectured that stars much more massive (five or ten times) than the sun undergo a much more violent collapse, and that the outer layers of the star are ejected into space in a supernova explosion, leaving behind a collapsed star called a neutron star. The properties of neutron stars were studied theoretically by the Russian physicist Lev Landau and independently by a group of American physicists, Robert Oppenheimer, G. Volkoff, and Harlan Snyder. They found that, in a neutron star, the atoms that make up ordinary matter are crushed completely. The electron clouds around the nuclei of the atoms are in effect crushed into the nuclei, where they combine with protons to make neutrons. Under ordinary conditions a neutron decays in a few minutes, back into a proton and an electron. In a neutron star, however, the extremely high densities and particle energies make the proton and electron immediately recombine into a neutron, so that almost all the matter in a neutron star remains in the form of neutrons. The collapse of the core of a star into a neutron star has no analogy on earth. It is as if a structure the size of the Empire State Building were to collapse to a heap one centimeter high. A sphere of neutron material the size of the period at the end of this sentence would weigh about twenty million pounds — about ten trillion times more than a lead sphere of the same size.

The calculations of Landau, Oppenheimer, Volkoff, and Snyder indicated that a neutron star should have a radius of between six and ten miles. As for the surface temperature, they did not venture to speculate; certainly these stars must have been extremely hot (billions of degrees) when they were formed, because of the heat generated in the catastrophic collapse. The question was, how long does it take a neutron star to cool down? Hong Yee Chiu of NASA's Goddard Space Center computed a surface temperature of about ten million degrees for a neutron star several hundred years old. Donald Morton of Princeton repeated the calculation and got a similar answer. These stars would evidently radiate enormous quantities of x-rays and little visible radiation. They appeared to be a neat solution to the puzzle of x-ray stars.

Confidence in the neutron-star hypothesis was bolstered by the fact that one of the sources of x-rays, the Crab Nebula, was known to be the remnant of a supernova explosion that had occurred nine hundred years before. Friedman's group at NRL developed an ingenious method for testing the neutron-star hypothesis for this x-ray source. On July 7, 1964, the moon would eclipse the Crab Nebula. By observing the x-ray emission from the Crab Nebula during this eclipse, they would be able to tell whether the x-rays were coming from a point-like star or from the nebula as a whole. If the x-rays came from a point-like source, the x-radiation would disappear suddenly as the moon eclipsed it. But if the x-rays came from the nebula as a whole, the observed x-radiation would fall off gradually as the moon occulted more and more of the nebula. It was very similar to the x-ray experiments Friedman's group had conducted during the solar eclipse seven years earlier. But it would be much more difficult because of the weakness of the source.

Since the moon eclipses the Crab Nebula only once every nine years, that July 7 would be the last chance of the decade. The group rushed to meet the deadline, even though the automatic control system for the Aerobee rocket was still in the development stage and had failed to function properly in six previous attempts. In an impressive and elegant experiment, they launched the rocket at precisely the right time, the automatic control system worked perfectly, and the x-ray detectors pointed steadily at the Crab Nebula for those few minutes of the decade during which the central regions of the nebula were hidden behind the moon.

The experiment was a success, but the results were disappointing. An analysis of the data showed that the x-ray emission did not drop off suddenly but declined steadily as the moon occulted the central parts of the nebula. At least 90 percent of the x-radiation was coming from the nebula at large. It was possible that 10 percent might come from a starlike object, but the NRL experiment could not prove this. In the race to meet the launch deadline, the group had had to keep electronics for the counters very simple. For this reason, they used count-rate meters that averaged the observed counts over intervals of time, rather than the single-photon counting devices used by the AS&E group. Had they been able to use single-photon counting devices, they would have detected a contribution of 5 to 10 percent of the x-radiation from a centrally located neutron star, and would have obtained the first strong evidence for the existence of neutron stars.

A few weeks after the NRL experiment, George Clark used a high-altitude balloon to measure the spectrum (the distribution of intensity with wavelength) of the high-energy x-radiation from the Crab Nebula. It was not consistent with the spectrum of radiation expected from the hot surface of a neutron star. Rocket observations of Sco X-1 by the AS&E and Lawrence Livermore groups led to a similar conclusion, namely that the x-radiation from Sco X-1 was not consistent with the hypothesis that Sco X-1 was a hot neutron star. Then theorists John Bahcall and Richard Wolf made a more accurate calculation of the expected surface temperature of a neutron star a few hundred to a thousand years old. They concluded that it should be well below ten million degrees. At such a temperature, a neutron star as far away from earth as the Crab Nebula (about five thousand light-years) would not have been detectable with the instruments then in use. It seemed unlikely that the x-rays from either the Crab Nebula or Sco X-1 were emitted by hot neutron stars.

What then *were* x-ray sources such as Sco X-1? They were clearly different from the Crab Nebula source, which was associated with a nebula that glowed brightly at visible and radio wavelengths. Nothing of the sort was detectable around Sco X-1 or any of the other x-ray stars. But were they truly stars? Was one of the many faint stars seen in the region the visible counterpart of Sco X-1? Or was Sco X-1 a cloud of hot gas around an invisible star or group of stars?

The AS&E group believed they could make a crucial test of these two hypotheses if they could greatly improve the angular resolution of the detectors. One way to do this, as Friedman's group had demonstrated, was to use the moon to eclipse the source. Obviously this was not a practical method for general use. In essence what was needed, in the absence of an x-ray telescope, was a detector that created its own eclipses.

This is done by means of collimators, that is, slats or wire grids placed over detectors. Minoru Oda suggested the arrangement for the grids. As the source of x-rays moves across the grids, it will produce a pattern of radiation that depends on the size of the source. To get a rough idea of how a collimator works, place your outstretched fingers in front of your eyes and move your fingers slowly across this page. Large blocks of words will be only partially hidden, but individual words will appear and disappear as your fingers occult them. In much the same way, it is possible to estimate the size of an x-ray source by using collimators in front of a detector.

Sophisticated collimators were developed at MIT by Minoru Oda, Hale Bradt, Gordon Garmire, and Giancarlo Spada. Two of these collimators were included in a rocket payload conceived by Herb Gursky at AS&E. With these instruments, the AS&E-MIT team hoped to pin down both the size and the location of Sco X-1. The rocket was launched from White Sands on March 8, 1966, and the experiment was a success. The data indicated that Sco X-1 was not an extended source like the Crab Nebula, and that in all likelihood it would appear in visible light as a starlike object. These results, along with a small region of the sky where the star must be located, were communicated by Oda to his colleagues at the Tokyo Observatory and by Giacconi to Allan Sandage and his colleagues at the Palomar Observatory. The Japanese astronomers found a faint, flickering blue star in the error box. The Palomar group confirmed their finding within a week, and Lockheed astronomers Hugh Johnson and C. B. Stephenson independently proposed the same star as the visible counterpart of Sco X-1 on the basis of its unusual appearance.

This star, which had not been previously named, became known simply as Sco X-1. It was observed to vary in brightness by a few percentage points over a period of a few minutes and by 50 percent over a day. Optical astronomers found this behavior reminiscent of old novae. Novae are stars that undergo large outbursts during which they become ten thousand times brighter for about a month.

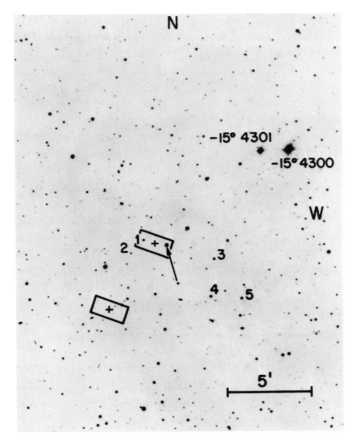

A Palomar Sky Survey print of the region containing Sco X-1. Two equally probable x-ray positions are marked by crosses surrounded by rectangles. The arrow marks the star identified with Sco X-1. Other comparison stars in the field are also marked. (A. Sandage et al., *Astrophysical Journal (Letters), 146* (1966), 316; by permission of the University of Chicago Press)

Unlike the much more energetic supernovae, for which the brightness increases by a billion times and the entire star is disrupted, novae apparently involve only the outer layers of the star. Also unlike supernovae, they are often observed to recur every twenty years or so.

In 1962 Robert Kraft of the Lick Observatory in California

brought together evidence that novae are members of double star systems, and suggested that the nova outburst is produced by the exchange of matter between a normal star and a white dwarf companion star. The cloud of dust and gas that collapsed to form the sun contained only enough matter to form one medium-sized star, with about a tenth of a percent left over for the planets. But in many cases, probably more than half, the protostellar cloud breaks up into two or more stars.

In some double star systems, the stars are so close together that they almost touch. The matter on the outer layers of these stars is torn between the gravitational force of the parent star and that of the companion star. As a result, the parent star can be distorted out of round. The same effect occurs on earth: the gravitational force of the moon pulls the earth slightly out of round, by about a foot; this distortion shows up as the high and low tides of the ocean. In a similar way, one star can raise tides on another star. In extreme cases, these tides can pull matter away from its parent star. The falling of the "stolen" matter onto the companion star, Kraft suggested, could produce nova outbursts.

Although there was evidence that in general old novae did not emit x-rays, the identification of the optical counterpart of Sco X-1 with a star having the characteristics of an old nova immediately led a number of astronomers to suggest that x-ray stars were members of double, or binary, star systems. Two Japanese astrophysicists, Satio Hayakawa and Masaru Matsuoka, had suggested in 1963 that binary star systems might be x-ray sources. Their model was different from Kraft's model for novae; rather than a white dwarf pulling matter from a nearby companion; they envisaged a very bright star that was losing mass because of internal processes. Many bright supergiant stars are observed to lose mass at a very rapid rate in a high-velocity "stellar wind" produced by the pressure of the star's own radiation. Hayakawa and Matsuoka reasoned that when the companion star plowed through this wind, the shock waves produced would generate x-rays.

In 1966 Bruno Rossi attended an International Astronomical Union symposium in Nordwijk, the Netherlands, where he reported on the optical identification by the teams at AS&E, MIT, Cal Tech, and Tokyo of Sco X-1 with a nova-like star. A group of interested astrophysicists convened a special meeting to discuss the implications of this identification. Oddly enough, they did not discuss Kraft's "overflow" nova model, but rather "wind" models of the type suggested by Hayakawa and Matsuoka. The major problem with this type of model, for Sco X-1 at least, was that there was obviously no supergiant start to supply the wind. Geoffrey Burbidge, recognizing this problem, suggested that perhaps an unseen neutron star was producing a wind that was interacting with the companion star.

Not long after this symposium, the Russian astrophysicist Iosef Shklovsky turned this idea around and came up with a model that was more akin to Kraft's model for novas. He postulated that matter was flowing from a more or less normal star onto a neutron star, and that the infall of the matter onto the neutron star was producing the x-rays. The general idea that x-rays could be produced by accretion of matter onto neutron stars was not new. Although many western scientists may have been unaware of it because of the time lag in receiving English translations of Russian work, accretion as a means of producing x-rays had been discussed in 1965 in a classic paper, "Relativistic Astrophysics," by Yakov Zeldovich and Igor Novikov of the USSR Space Research Institute in Moscow. Because of the neutron star's highly collapsed state, its gravitational field is tremendously strong. Matter falling toward a neutron star will be accelerated to high energies. As the matter interacts with the atmosphere of the neutron star, much of its energy will be radiated away at x-ray wavelengths. This was an attractive model because it used an efficient means of generating energy, namely the gravitational potential energy. All that was needed was a source of matter and a neutron star.

Soon after Shklovsky's model was published, Burbidge and Kevin Prendergast published a similar model in which matter from a normal star accreted onto a white dwarf star. Alistair Cameron and Michael Mock presented a similar model. In their paper, Cameron and Mock criticized Shklovsky's neutron-star model, showing that it would produce temperatures far in excess of what were observed. But the white dwarf models were also subject to the criticism, namely that white dwarfs in binary systems appeared to be far more common than x-ray stars. What special conditions were needed to produce x-ray stars?

A clear prediction of the binary-star model, whether it involved neutron stars or white dwarfs, was that it should exhibit periodic variability as the stars moved in their orbits around each other. For example, the source might show evidence of eclipses as one star moved in front of the other. The difficulty of testing this model with x-ray observations was obvious. Rocket flights lasted about five minutes, and high-altitude balloon flights a few hours, whereas the binary stars were likely to take days to complete their orbits. There was evidence from rocket and balloon flights of variability in certain x-ray stars. For example, Walter Lewin and his co-workers at MIT had used an x-ray detector carried by a balloon to observe an x-ray flare on Sco X-1 that lasted ten minutes. The x-ray emission from Cygnus X-1 had been observed to vary by as much as a factor of three from one flight to the next. The NRL group suggested that this variability might be caused by eclipses in a binary system. The prevailing opinion, though, was that expressed by Philip Morrison of MIT in 1967 in a review of the status of x-ray astronomy: "it is more probable that the systematic errors of various experiments have merely simulated changing sources." The final answer on the variability of x-ray stars would not be given until a detector was put aboard a satellite.

The first opportunity to fly an x-ray detector on a satellite met with singularly bad luck. In 1966 NASA asked the Lockheed group to put an x-ray detector on the first Orbiting Astronomical Observatory. But misfortune continued to plague the Lockheed group. Immediately after achieving orbit, the spacecraft suffered electrical malfunctions, resulting in the failure of all the experiments on board. Two other satellites carrying x-ray detectors were launched the next year. Larry Peterson of the University of California at San Diego put a small high-energy x-ray detector of the type flown in balloons on NASA's third Orbiting Solar Observatory, and a Los Alamos group led by Jerry Conner had small detectors aboard a Vela satellite launched in the same year. But these missions were not dedicated to looking at sources outside the solar system. The Orbiting solar Observatory was primarily for solar observations, and the Vela satellite was primarily used to detect clandestine nuclear explosions in the upper atmosphere. As a result, the coverage of sources outside the solar system was limited, and the experiments did not provide much information on the nature of x-ray stars.

Optical observations held out the best hope. Even if no eclipses occurred, it seemed likely that the shifting radiation pattern produced by the motion of the stars could be detected, thus verifying the binary-star model. This follows from the Doppler shift, according to which the frequency of radiation is shifted by the motion of the source of the radiation relative to the observer. The frequency increases if the source is moving toward you and decreases if it is moving away. This effect applies to any type of wave motion; a familiar example is the rise in pitch of a train whistle as it approaches you and the decrease in pitch as the train rushes away from you. This effect is observed in many binary star systems; the systematic shift up and down in frequency can be used to trace the motions of the stars emitting the radiation and to determine the period of the binary orbit. When this technique was applied to Sco X-1, it yielded confusing results. Doppler shifts were observed, indicating motion, but no pattern in the shifts was apparent. Perhaps the observers were seeing streams of matter being ejected from the parent star, or falling onto a neutron star. Perhaps not.

In the hope of finding a "cleaner" object to study, the AS&E group determined to fix accurate locations for other x-ray sources, in order to look for their optical counterparts. The constellation of Cygnus, with three strong sources, was chosen for this attempt. In October 1966 the group succeeded in obtaining accurate locations for three sources, Cygnus X-1, Cygnus X-2, and Cygnus X-3. Of the many candidate stars that appeared in the x-ray error boxes, only those which resembled the optical counterpart of Sco X-1 were considered. This seemed to be an eminently reasonable assumption based on the one known case. A faint blue star was identified with Cygnus X-2. No suitable stars could be found for Cygnus X-1 and Cygnus X-3. We now know that the restriction to stars similar in visible appearance to Sco X-1 was a mistake. Cygnus X-1 has since been identified with a bright blue supergiant star that shows clear evidence of binary motion. Had the scientists not had their minds on Sco X-1, the binary hypothesis might have been verified years earlier than it was.

Cygnus X-2, meanwhile, turned out to be much like Sco X-1; there were Doppler shifts, perhaps even some evidence of binary motion, if you were a believer. But not if you were a skeptic. So the binary-star hypothesis, while still popular, remained just that, a hypothesis, with no facts to back it up. Then in 1968 came another discovery, this time from radio astronomy, that further confused the issue.

In late 1967, British radio astronomers Jocelyn Bell and Anthony Hewish discovered a source of radio waves that exhibits extremely regular pulsations, with a period close to one second. This object blinked on and off with the precision of a clock that loses only one second in a million years. For a while these objects were referred to as LGM's, for Little Green Men, because of the possibility, seriously entertained, that they might be signals from an extraterrestrial civilization. After more of these objects were discovered they became known by the less colorful name, "pulsars." What were they? Only the pulsation or rotation of a star could maintain such a regular period of radio pulsation, and only a collapsed star such as a white dwarf or a neutron star could rotate or pulsate as rapidly as once per second. Thomas Gold of Cornell University suggested that pulsars are rapidly rotating neutron stars. He predicted that, since a neutron star is formed as a consequence of the same event that leads to a supernova star, namely the collapse of a massive star, pulsars should be found in the remnants of supernovae, the Crab Nebula being the primary example. He further postulated that the pulsar was deriving its energy from the rapid rotation, and thus that it should be gradually slowing down. This might show up as a very small, but detectable, increase in the observed period of the radio pulses.

Within months, groups at the National Radio Astronomy Observatory in Green Bank, West Virginia, and the Arecibo Radio Observatory in Puerto Rico provided striking confirmation of Gold's predictions. At Green Bank, David Staelin and Edward Reifenstein III discovered a pulsar in the vicinity of the Crab Nebula. The Arecibo group, led by J. M. Comella, pinpointed the pulsar's location near the center of the nebula and made a detailed study of the pulse rate. It was producing radio pulses at the rate of thirty per second, and the time between pulses was increasing very, very gradually, by only a few tenths of a percent per year. Neutron stars, which for thirty years had, in the words of Zeldovich, "existed only at the tip of the theoretician's pen," had finally left the "world of paper" to become part of our "sensible" world.

Did the Crab pulsar also emit pulses of x-rays? Friedman's NRL group, with Gilbert Fritz taking the lead, rushed to be the first to find out. NRL won this one, when on March 13, 1969, they detected pulsed x-ray emission from the Crab. The interval between pulses coincided exactly with that measured at radio frequencies. A few weeks later an MIT rocket group led by Hale Bradt confirmed their results. A few weeks after that, a Rice University group led by Robert Haymes reanalyzed high-energy x-ray data from a 1967 balloon flight and showed that the pulsations extended to energies ten times greater than those detected by the NRL and MIT groups. The x-ray results showed that the pulsar was radiating most of its energy at x-ray wavelengths, a discovery that the group had just missed making more than three years earlier.

The rapid rotation of a neutron star is a natural consequence of the way in which neutron stars are formed. When an object collapses, any rotational motion it has is greatly amplified. A familiar example is the figure skater who speeds up her rotation by pulling in her arms. A neutron star is formed when the core of a star collapses to a diameter only one-ten-thousandth or less of its original diameter. Because of the amplification of rotational motion by collapse, the stellar core may be rotating a hundred million times faster than before the collapse by the time it forms a neutron star.

Another consequence of collapse is that the magnetic fields tied to the core of the star are amplified by similar very large factors. The rapidly rotating magnetic field of a neutron star can produce differences in electrical potential of trillions of volts. These voltages pull particles off the surface of the neutron star and accelerate them to enormous energies. This happens mainly near the magnetic poles of the star, producing two diametrically opposite beams of high-energy particles. These particles produce beams of radiation, which, because the star is whirling around (thirty times a second in the case of the Crab pulsar), show up as a sequence of regular pulses.

The discovery of pulsars proved the existence of neutron stars. It also showed the importance of a source of energy that had been overlooked, namely, the energy of the rotation of the neutron star. The Crab Nebula is one of the most striking manifestations of the power of this energy source. The Crab Nebula was produced by a supernova explosion that was observed in 1054 A.D. by Chinese, Japanese, Arabian, and possibly North American Indian astronomers. Today the nebula is observed as a cloud that stretches across several light-years and produces intense radiation from radio through gamma-ray wavelengths. A careful study of the characteristics of the radiation from this cloud shows that it comes from very high-energy electrons interacting with the magnetic field of the cloud.

The existence of this cloud of high-energy particles posed a major puzzle for astrophysicists. The particles producing the x- and gamma radiation should radiate away most of their energy in a few years or less, yet the cloud has been around for a thousand years. How is the supply of high-energy particles replenished? One prescient suggestion was made in 1967 by Franco Pacini, an Italian astrophysicist. He proposed that a rotating neutron star with a large magnetic field would radiate large amounts of electromagnetic energy that could somehow energize the nebula.

The discovery of the pulsar in the center of the Crab Nebula confirmed Pacini's suggestion and solved the problem almost overnight. The neutron star there was rotating thirty times a second, generating electromagnetic fields that in turn produced the high-energy particles in the cloud as a whole. Peter Goldreich and William Julian of UCLA and James Gunn and Jeremiah Ostriker of Princeton showed in two separate papers how this could be understood in detail.

The energy for the particles comes ultimately from the rotational energy of the neutron star. If this is true, then the neutron star should be slowing down by an amount that can be estimated from the amount of energy lost in the form of radiation. In a clear confirmation that occurs all too seldom in astrophysics, it was shown that this is indeed the case. Observations of radio, optical, and x-ray pulses showed that the pulses were gradually getting farther apart. The neutron star is slowly spinning down, with each successive spin taking about a trillionth of a second longer. The rotational energy of the star is gradually decreasing by just the amount required to account for the energy radiated by the neutron star.

The stunning success story of the Crab pulsar prompted theorists to propose a new class of models for x-ray stars such as Sco X-1. In these models the x-ray stars, like the Crab pulsar, were energized by a highly magnetized, rotating neutron star. A major problem was that many of the x-ray stars had no glowing nebula around them; this was explained by assuming that these x-ray stars were much older than the Crab pulsar, or that they had different characteristics, such as a much stronger magnetic field or a higher density of surrounding gas, so that a nebula was never formed. According to these models, the x-ray stars might be expected to exhibit rapid pulsations or to show the effects of the very strong magnetic fields. But none of the strong x-ray sources showed any evidence of periodic pulsations on a scale of seconds. Tests for rapid pulsations, on a scale of milliseconds, could be more practically carried out from satellites.

The decade of the 1960s had been one of the most exciting in the history of astronomy. It had seen the discovery of quasars, the microwave background radiation, the x-ray background radiation, x-ray stars, and pulsars. All these discoveries had come about as a result of radical improvements in observational techniques. It was, therefore, with understandable excitement that x-ray astronomers looked forward to the launch of the first satellite devoted exclusively to x-ray astronomy. It would bring about a major leap forward in observational capability. It held the promise of making new discoveries and of solving the riddle of the x-ray stars.

7

Uhuru: Neutron Stars and Black Holes

Once the first cosmic x-ray sources had been discovered, it was inevitable that satellites would be used to study them. The only question was when. Giacconi and his colleagues at AS&E addressed this question in the fall of 1963. They were seeking support for rocket flights to continue the broad surveys of the sky for cosmic x-ray sources that had been initiated under the sponsorship of the Air Force. At the time only the existence of Sco X-1 and the Crab Nebula x-ray source had been established. Nevertheless, they were convinced that a rich new field was opening up. Through a series of long discussions with Herb Gursky and others, Giacconi developed a long-range plan. Undeterred by the embryonic state of the field, and emboldened by the conviction that the overall concept was sound, he presented to Nancy Roman, chief of the Astronomy Branch of NASA, an ambitious program for the development of x-ray astronomy.

The program had five phases. The first was the continuation of the rocket program of surveying the sky for new sources, studying selected sources in detail, and testing new instrumentation. The second phase would involve an experiment on the fourth Orbiting Solar Observatory. These observatories were large, spinning, wheel-like spacecraft supporting a section that pointed toward the sun; secondary experiments were positioned in the rim of the wheel, and could scan the sky as the wheel rotated. The third phase of the program would involve an "x-ray explorer" satellite, a satellite designed especially for the purpose of doing x-ray astronomy. The fourth phase would be a pointed spacecraft of modest size carrying a focusing x-ray telescope. Finally, there would be a large orbiting x-ray observatory carrying a 1.2-meter x-ray telescope. The large x-ray telescope was projected to fly in the late sixties.

The hope, of course, was that NASA would endorse the entire plan. Realistically, Giacconi and his colleagues expected that the rocket program would be approved, and that a space on the Orbiting Solar Observatory might be possible. The other parts of the program were developments that would ultimately have to occur if the field were to mature into a full-fledged partner with optical and radio astronomy. It could not hurt to suggest them, and it might even help, even if only by stimulating thinking about x-ray satellite observatories within the halls of NASA.

The rocket program was approved as expected, and, much to the surprise and delight of the AS&E group, Roman expressed an interest in an x-ray explorer satellite. She asked for more details. Within months the group prepared a detailed proposal, "An X-Ray Explorer to Survey Galactic and Extragalactic Sources"; on April 8, 1964, they submitted it to NASA. The satellite's mission would be to make a detailed survey of the sky for x-ray sources, in the hope of finding new sources and improving our understanding of the known sources. In the proposal, the AS&E group maintained that several requirements had to be satisfied in order to optimize the scientific return from the mission. One requirement was that the satellite should spin very slowly, so the detector would scan very slowly over a small angle in the sky. As the detector scanned across an x-ray source, the slow rate of scanning would keep the source in its field of view for a long time. This would make it possible to discover new, weak x-ray sources and to determine the accurate positions of these and other, already known sources. Knowing the positions would make it possible, the group hoped, to identify the visible counterparts of these sources.

Another requirement was that the satellite should be launched from the equator. The earth rotates about 13 percent faster at the equator than at Cape Canaveral in Florida or Vandenberg Air Force Base in California, so a rocket launched at the equator gets an extra boost from the earth's rotation. This meant that a slightly larger payload could be launched. It also meant that the satellite would avoid those troublesome regions in the earth's magnetic field where the concentration of trapped charged particles is high. These and other requirements led the AS&E group to propose the construction of an X-Ray Explorer spacecraft especially for this mission. The mission was to last eighteen months, beginning with a launch on December 5, 1965.

A NASA reviewing committee of astronomers, chaired by Roman, endorsed the mission. Then came the question of implementation. NASA decided that AS&E could be entrusted with the construction of the scientific payload, but not with the construction of the spacecraft. That would have to be done elsewhere, and a NASA center would have to manage the project.

NASA is a complex organization with the broadly defined mission of coordinating and directing aeronautical and space research in the United States. It accomplishes this mission in two ways: first, it contracts with private companies, universities, university laboratories, and other government agencies to do work of one type or another; second, it maintains a number of centers. Examples of these are the Johnson Manned Space Flight Center in Houston, which is responsible for the development and operation of manned space flights, the John F. Kennedy Space Center at Cape Canaveral, which is responsible for preflight testing, checking, and launching of space vehicles, and the Goddard Space Flight Center in Greenbelt, Maryland, which is responsible for the earth-orbiting satellites and is the headquarters for the NASA network of tracking stations.

As new programs come up, the responsibility for carrying them out is delegated by NASA headquarters to one of these centers. Realistically, a program can proceed only if the staff of the center that is to execute it takes an active interest in it. It is very difficult for a center to become very excited about a unique small program that is of interest only to a small group of scientists outside of NASA. The X-Ray Explorer project met with a lukewarm response from NASA center managers at first. Fortunately, John Naugle, then NASA's associate administrator for science and applications, gave unwavering support to this mission.

Finally, in late 1966, two and a half years after the submission of the detailed proposal and a year after the proposed launch date, NASA decided that the X-Ray Explorer was such a good idea that it should be generalized. NASA would initiate a program of Small Astronomy Satellites, of which the X-Ray Explorer would be the first. The program would be directed by Goddard Space Flight Center, with Marjory Townsend as the program manager. The Applied Physics Laboratory of Johns Hopkins University was chosen to design and build the spacecraft. In December 1966, the AS&E

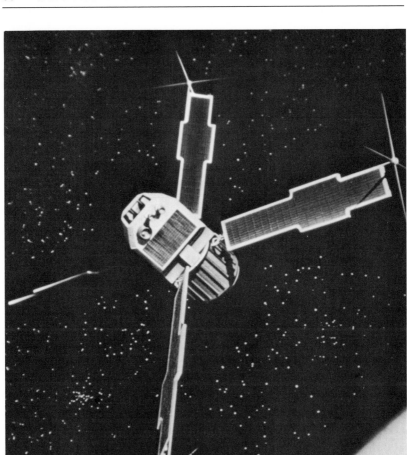

An artist's rendering of the *Uhuru* (X-Ray Explorer) satellite. (American Science and Engineering)

group submitted a proposal that described in great detail how the mission would be carried out.

Giacconi's basic philosophy with regard to the design of the spacecraft was straightforward: make it simple and make it reliable. The design took advantage of the great advances in the technology of constructing very thin beryllium windows for the proportional counters. These counters had been developed by John Waters of AS&E for an Apollo mission payload that was eventually canceled. There had also been progress in reducing the background noise; a new technique developed by the Leicester group in England had been modified for use in x-ray astronomy by Paul Gorenstein of AS&E.

In the three years that had passed since the X-Ray Explorer was first proposed, x-ray astronomy had undergone rapid growth, both in knowledge about individual sources and in techniques. About twenty sources were known, the lunar occultation experiment on the Crab Nebula had been performed, and Sco X-1 had been optically identified; the slow scan and collimator techniques had been developed, along with the improved proportional counters and methods of rejecting background noise. These developments were the subject of almost continual discussions among the scientists and the engineering support staff at AS&E. The need to maximize scientific returns on a fixed budget forced the group to simplify the original concepts. As the engineers contributed the latest in technological developments and feasibility analysis, the group made careful tradeoffs between added complication and added reliability. It was in this atmosphere that the elegant simplicity of the X-Ray Explorer's design evolved.

American Science and Engineering was in those days an unusual institution: a small company that derived a large part of its revenues from doing pure science. To what extent did these peculiar institutional conditions facilitate the building of the X-Ray Explorer in particular, and the development of x-ray astronomy in general? When compared with an academic institution, AS&E offered, in our opinion, several advantages. First of all, there were no distractions, such as faculty meetings, courses to teach, and other academic duties, from research and the management of research. Secondly, there were no limits on the expansion of personnel and facilities, provided only that funds could be obtained. This was related to the third advantage, namely the ability to hire and fire at will. There was no tenure system. Management errors and technical

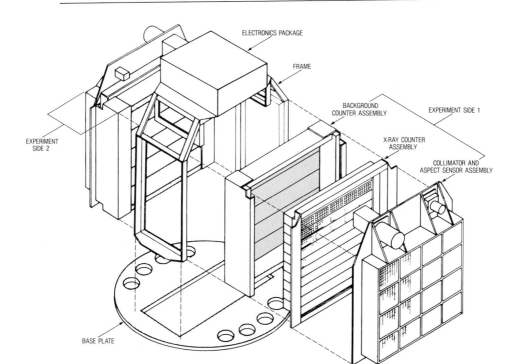

An exploded schematic of the instrumentation in *Uhuru.* (American Science and Engineering)

incompetence could be and were quickly remedied; new skills were acquired very rapidly through the expansion of the work force. Because AS&E was interested in growth in commercial and defense-related research as well as in pure science, it expanded rapidly in the 1960s, and technical competence could be drawn from a constantly expanding pool. This enabled the company to carry out several development programs simultaneously; for example, proportional-counter technology could be developed at the same time as x-ray telescopes. It was not necessary to wait until one project was finished to start another one. Finally, AS&E gave its scientists the freedom to take risks and the benefit of management investments in capital, equipment, and seed money with which to pursue new directions of research.

Because of the nature of AS&E, Giacconi was given essentially free rein to develop the space science program from the day he was hired. At the outset, he had six months to think about little else. This was partly fortuitous; he had no security clearance, so he could be only peripherally involved in work for the Department of Defense. He was able to draw upon a talented pool of consultants from MIT, such as George Clark and Stan Olbert, and he was able to use the available shops to start to work on prototypes of telescopes almost immediately. As chief of the Division of Space Research, which he had created, he exercised very direct control over the project and the staff. In the early 1960s AS&E expanded its research program to pursue research in six different areas of space science: experiments on trapped radiation in the Van Allen belts; a rocket program in x-ray astronomy; an experiment connected to the Apollo program; the laboratory development of x-ray telescopes; the use of solar x-ray telescopes; and experiments for the Orbiting Solar Observatory.

Throughout this period, the AS&E group developed a strong support capability in systems, electrical, and mechanical engineering. Each Ph.D. scientist who was carrying out an experiment had a crew of three engineers and one or two technicians. With this support, small groups could carry out experiments quickly and successfully. Laboratory space expanded from one to eight old milk-truck garages. Technology improved rapidly; Waters, in addition to his work on proportional counters, introduced computerized ground-support equipment. Large numbers of tests could be conducted in a very short time and data from these tests could be assessed as they were being conducted. This technology became essential, and was used in the X-Ray Explorer program.

Don Frieklund, a systems and aeronautical engineer, was instrumental in developing the design of the X-Ray Explorer. He came to AS&E in 1963 from RCA, where he had been building the TIROS weather satellites. He brought his talent for mechanical design to bear on the X-Ray Explorer and other AS&E programs. The satellite and rocket instruments developed an organic look, in which the skin of the instrument became like an exoskeleton and supporting structure, in contrast to the old design wherein the rockets had skeletons on which we would hang instruments. The new design made the payloads lighter and more compact, and permitted larger detection areas for a given payload weight.

Although AS&E's first Apollo experiment was not flown, the

experience was valuable. The establishment of quality control to meet the requirements of manned flights, and the use of simple, redundant components made our electronic state of the art reliable and failure-free. In general, good engineering led to a large payoff in scientific returns. The existence of a strong engineering group encouraged the scientists to think of sophisticated and ambitious experiments, because they knew they could be carried out in practice. Examples are the development of x-ray telescopes for solar x-ray astronomy and the collimator experiment of 1967. It was because of these scientific and technological accomplishments that the AS&E group was able to assume a position of leadership for the large x-ray telescope mission in the 1970s.

In summary, institutional support was essential to the success of the AS&E group. Administratively and materially, the company encouraged the scientists to succeed. For example, Giacconi was able to convince the board of directors of AS&E to use $100,000 of corporate funds to complete the testing of the satellite. The Space Research Division was allowed to grow according to the dictates of logic, rather than by detached management decisions or artificial limitations. It was in this spirit that Giacconi hired Wallace Tucker in 1969 to head a small theoretical astrophysics group. Though not directly involved in the building of the experiment, Tucker and his group supported the effort by making theoretical calculations and by collaborating on technical papers. Perhaps more important, through informal discussions and formal seminars the theoretical group provided astrophysical background information for the experimentalists, most of whom had been trained in physics rather than astronomy, and they helped the experimentalists to maintain an awareness of recent theoretical developments in astronomy. This prepared the AS&E scientists to interpret the data from their experiments and assisted them in long-range planning, in which Tucker was more directly involved.

The strong link between the engineers and the scientists left the scientists free to think about scientific problems without, at least initially, having to worry about engineering difficulties, and it led to simple, inexpensive, and reliable designs. An intimate part of the design of the X-Ray Explorer was the data handling system. The AS&E group wanted to be able to understand new discoveries quickly enough so that they could modify the observational program while the satellite was still operational. The satellite was expected to last only six months, so unprecedented speed was required in the collection and digestion of the essential features of the data. Harvey Tananbaum and Edwin Kellogg were largely responsible for the painstaking attention given to the design and implementation of star sensors, which would fix on known reference stars and send back information about where the satellite was pointing. Kellogg and Tananbaum also developed the data-processing electronics so that they were compatible with a fully automatic system for checking the performance of the spacecraft. So they would understand exactly what was happening on the spacecraft as it collected and transmitted data, they designed and implemented a highly automated software system for decoding the signals from the spacecraft into information about the performance of the spacecraft and what it was observing. The staff of Goddard Space Flight Center cooperated fully in the endeavor to speed up the collection and analysis of data. They arranged to have data transmitted from Goddard to AS&E every day over telephone lines. This ability to monitor the data daily proved to be crucial in solving the riddle of the x-ray stars.

When completed, the X-Ray Explorer weighed 64 kilograms. It had independent detection systems pointing in opposite directions. The detection system consisted of a set of beryllium-window proportional counters with collimators in front. One collimator had a field of view of .5 degrees by 5 degrees (about the same angular area as a pencil held at arm's length); the other detector had a field of view of 5 degrees by 5 degrees (roughly the angular area of this book held at arm's length). The effective area for collecting x-rays was about 700 square centimeters, approximately the size of a 9x12-inch photograph. The gas filling the counters was mostly argon, with 10 percent carbon dioxide and a trace of helium for the detection of leaks.

The satellite spun slowly, making one revolution every 12 minutes. A magnetic torquing system, which reacted against the earth's magnetic field, could be used to slow the spin to less than one revolution per hour. During normal operations the satellite spin axis pointed one direction in the sky for about a day, and then the magnetic torquing system was used to point it to a new position. In this way the satellite scanned the whole sky. In analyzing the data, it was important to know exactly where the satellite was pointing, since a variety of perturbing forces, such as small variations in the gravitational field and the interaction of the earth's magnetic field with the satellite, caused the spin axis to drift continually. The task of understanding this drift, and hence the position or aspect of the satellite, fell to Ethan Schreier, who became known as "Mr. Aspect."

The data were to be recorded by accumulating counts for a fixed time. The shortest accumulation time was one-tenth of a second. After the pulsar in the Crab Nebula, with its period of 33 milliseconds, was discovered in 1968, AS&E requested about $200,000 from NASA to modify the data-recording system so that rapid variation on a time scale of tens of milliseconds could be studied. The request was denied on the recommendation of Marjory Townsend, who preferred to use the funds for further safeguards of the electronics against temperature variations in orbit. In retrospect, this was a misplaced concern, because the temperature in the spacecraft never varied more than one degree. This limitation on the response of the counters to rapid variations made it impossible to study the Crab pulsar and greatly complicated the analysis of the data from the peculiar source Cygnus X-1. Fortunately, though, the accumulation time of one–tenth of a second did prove adequate for a major breakthrough in understanding the nature of the x-ray stars.

In November of 1970 the X-Ray Explorer experiment payload was shipped to Kenya in preparation for the launch. The launch was to take place from the San Marco Platform, an old oil-drilling rig three miles offshore from a small Kenyan village where food was cooked over open fires and electricity and plumbing had yet to arrive. The oil rig had been modified for rocket launches by the Italian Space Agency. Tananbaum and two AS&E engineers, Dick Goddard and Stan Mickiewiz, went to Kenya a month before launch for the prelaunch checkouts, along with teams from the Applied Physics Laboratory (APL) and NASA. At 7 A.M. every day for a month they made the three-mile boat trip out to the platform, to test the spacecraft and the experiment to make sure that all was well. Giacconi and Townsend joined them in Kenya about two weeks before launch.

Everything went smoothly until two days before launch. Then one of the batteries that was to supply power to the experiment began to give anomalous readings. Was the battery failing? Or were the readings spurious, a not uncommon occurrence? Should the instrument package be pulled out of the spacecraft and a new battery, one that had not gone through the rigorous checkout proce-

Preflight testing of the *Uhuru* satellite on the San Marco launch platform. (Italian Space Agency)

dure, be installed? The APL group responsible for the batteries opted out of the decisionmaking process. They took a boat back to shore and left the decision to Marjory Townsend. She decided to replace the battery.

The launch was rescheduled for dawn on December 12, 1970. But delays of one sort or another kept halting the countdown. Giacconi spent the night before launch on the platform itself, where one could catch a few minutes' sleep on the wet steel decking. During this hectic prelaunch period, some of the Italian crew literally lent their American colleagues the shirts off their backs. As the sun rose in the sky, so did the anxiety level of the scientists and engineers. Was the spacecraft heating up so much that the electronics would be damaged? To cancel the launch at this point would mean costly delays, and with delay there was always the chance that something else might go wrong. Yet to launch might be risky. Finally, shortly after noon, Townsend and Giacconi decided to launch. As the scien-

Giacconi with the Italian director of the San Marco facility, General Broglio. (Italian Space Agency)

tists and engineers, along with the Kenyan villagers, watched on an exceptionally clear day, the Scout rocket flawlessly carried the satellite into a five-hundred-mile-high orbit over the equator. Because December 12 was Kenya's Independence Day, the satellite was nicknamed *Uhuru*, the Swahili word for "freedom," in appreciation of the cooperation of the Kenyan people.

After launch, Giacconi was anxious to find out if the instrument worked as designed. He convinced Townsend to turn on the high voltage so they could obtain a first glimpse of the x-ray sky when *Uhuru* made its first passage over Kenya. As the satellite circled the earth, they rushed by rubber boat across the three miles of water to the base camp. They reached the control van in time to turn on the detectors. As soon as they confirmed that the detectors were working perfectly, they shut them off again, hoping that Spacecraft Control at Goddard Space Flight Center would ignore this breach of procedure.

Giacconi and Tananbaum hurried back to Cambridge, where they joined the rest of the AS&E group, eager to get the first detailed view

Uhuru ready for launch. (NASA)

The launch of *Uhuru* on December 12, 1970. (NASA)

of the x-ray sky, a view that everyone hoped would be breathtaking. They were not disappointed.

One of the first objects to be studied was Cygnus X-1. This x-ray star was a peculiar object among peculiar objects. Despite a fairly accurately pinpointed location, it had not been identified with a visible counterpart, as had a weaker nearby source, Cygnus X-2. Its x-ray spectrum resembled that of the Crab Nebula x-ray source more than those of the other x-ray stars, yet there was no evidence of an extended nebula around Cygnus X-1. Then there was the question of its variability. Several different groups had measured the intensity of x-rays coming from Cygnus X-1 on several different occasions. Sometimes the results between two groups agreed, sometimes they did not. There were only two possible explanations. One was that some of the observations had been misinterpreted, so that systematic errors had merely simulated a changing source. In other words, someone was wrong. The other possibility was that the x-ray luminosity of Cygnus X-1 was changing with time. If this was so, then it was perfectly natural that different observations made at different times yielded different results. In other words, maybe everyone was right. With an eye toward resolving this controversy and obtaining a more accurate position so that a visible counterpart could be located, the group scheduled frequent and immediate observations of Cygnus X-1. It was observed on December 21, 1970, just nine days after launch, again on December 27, and on January 4, 1971. From this work came a publication by Tananbaum and his colleagues at AS&E, giving an improved location of Cygnus and verifying that Cygnus X-1 was indeed a variable source on a time scale of weeks.

The long-term variability of a source over time could be measured by comparing successive passes over the source separated by 12 minutes, the rotational period of the satellite. It occurred to Giacconi and Gursky that they could study shorter time variability within a single pass (which lasted 0.25 seconds and 2.5 seconds for the two different detectors) by comparing the observed distribution of counts with the expected response for a steady source. A steady source would appear, within statistical uncertainties, as a smooth triangular distribution of counts. That is, the count rate would smoothly increase as the source entered the field of view of the detector, peak when it was in the center of the field of view, and then smoothly decline as the source moved out of view. This is analogous to the amount of sunlight that shines in a window as the sun moves across the field of view of the window during the course of the day. A variable source, in contrast, would show departures from this pattern: it would be as if clouds moved across the sun during the day; the intensity of the sunlight coming through the window would vary not smoothly but erratically. This technique, which had not been thought of before launch, became the standard way to study the short-term variability of sources. Minoru Oda, who had taken a leave from the Institute of Space Science in Tokyo to visit AS&E for a few months, took a more detailed look at the Cygnus X-1 data to obtain a better understanding of its short-term variability. Oda quickly established that the intensity of Cygnus was varying, possibly rapidly and periodically. Was Cygnus X-1 a pulsar? Giacconi, Gorenstein, Tananbaum, and Schreier joined in the analysis of the data. On some scans through the source, it appeared to pulse regularly with a certain period; on other scans it seemed that the pulse period had changed; still other scans showed no evidence of any pulse period at all.

We both participated in impromptu discussions with other scientists at AS&E and with Bruno Rossi and Bruno Coppi at MIT as we tried to make sense of the data. A working hypothesis was that the source was pulsing with a period that was less than the data-accumulation time of a tenth of a second, so that it was blurred out. It was not possible to determine a unique period for the hypothetical x-ray pulses, but a pulse period of 73 milliseconds seemed consistent with the data. This led to a discussion of models based on rapidly rotating, highly magnetized neutron stars, by analogy with the Crab Nebula pulsar. An obvious difficulty in applying such models to Cygnus X-1 was the absence of radio emission and an extended nebula around the Cygnus X-1's x-ray star. Alternate models were proposed, such as a massive collapsed star stabilized by rapid rotation.

A more intriguing possibility was that Cygnus X-1 was a black hole. The theory of stellar evolution predicts that at a certain critical time, when the core of a star has used up its nuclear fuel, the core will collapse. If the star is about the same mass as the sun, it will turn into a white dwarf star. If the star is somewhat more massive, it may undergo a supernova explosion that leaves behind a neutron star, in which gravitational forces are held in check by nuclear forces. But if the collapsing core of the star has a mass greater than about three times that of the sun, gravitational forces overwhelm nuclear forces and the core collapses. Since nuclear forces are the strongest repulsive forces known, nothing can stop the continued collapse of the core. The star implodes, that is, it collapses to form a warp or a sort of bottomless pit in space, which is called a black hole.

John Wheeler gave black holes their name, but it was Robert Oppenheimer and Harlan Snyder who first studied their properties, and published their findings in a classic paper in 1939. They showed that, because of the intense gravitational field near a black hole, nothing can escape from it, not even light or other forms of electromagnetic radiation. In the words of Oppenheimer and Snyder, "the star thus tends to close itself off from any communication with a distant observer; only its gravitational field exists."

A black hole does not have a surface in the usual sense of the word. According to physics as we know it, the matter that falls into a black hole never escapes. What does it mean to fall inside an object that has no surface? Essentially, it means to fall inside the gravitational radius; once a particle gets too close to a black hole, the gravitational forces are so strong that the particle cannot escape, no matter how much energy it is given. The point where this happens is called the gravitational radius. For a black hole containing about ten times the mass of the sun, the gravitational radius is about ten kilometers. Matter falling inside the gravitational radius collapses forever. Because of the extreme nature of black holes, many physicists in 1971 were skeptical of their existence. They believed that as-yet-unknown effects would intervene to prevent the formation of black holes. Other physicists, such as Kip Thorne of the California Institute of Technology and Wheeler and Remo Ruffini of Princeton University, found the concept captivating, and their work and writing did much to popularize the black holes and to make them seem possible. At the forefront of research on black holes were Yakov Zeldovich and Igor Novikov of the USSR Space Research Institute, who suggested in the mid-1960s that black holes might be detected as x-ray sources.

The mechanism for the production of x-rays by a black hole is similar to the neutron-star-accretion model. As matter swirls into the gravitational maelstrom of a black hole, it accelerates under the action of the ever-increasing gravitational forces. This increased energy of motion is changed into heat energy by friction. Near the gravitational radius, temperatures may range from tens to hundreds of millions of degrees. At these temperatures, x-radiation and gamma radiation are produced. A black hole supplied with a stream of matter by a nearby companion star in a close binary system should therefore be a strong source of x-rays.

We would not expect the x-radiation from a stream of matter falling toward a black hole to show any regular or periodic behavior, since it has no well-defined surface. Instead, an accreting black hole might show erratic short-term variability on a scale of a tenth of a second or even a thousandth of a second as successive streams or blobs of matter fall toward the black hole.

The AS&E group seriously considered the possibility that the x-radiation from Cygnus X-1 is produced by the infall of matter toward a black hole. But if the 73-millisecond periodicity was real, it could not be a black hole. In a paper summarizing their observations, Oda and his colleagues avoided any detailed discussion of models in view of the uncertainties in the data, and emphasized that "we cannot exclude the possibility that the period is a multiple of 73 milliseconds or that a much more complex pattern of emission from the source may be degraded by our coarse time resolution . . . It is essential that these results be confirmed by continuous observations made with fine time resolution, such as should be possible with balloons or sounding rockets." Nevertheless, Giacconi, who was convinced from the observed behavior of Cygnus X-1 that it was different from all other x-ray sources the group had observed, insisted, as a coauthor, that the paper at least mention the possibility that Cygnus X-1 was "a collapsed object such as a neutron star or a black hole."

While x-ray groups at MIT and Goddard Space Flight Center planned sounding-rocket flights to search for the suggested 73-millisecond period, the AS&E group continued their analysis of data from Cygnus X-1. They also began to use the observational and analytical techniques developed for Cygnus X-1 to search other sources for evidence of periodic variations.

They soon found it. One morning in January, Schreier, Giacconi, and Tananbaum began to analyze the computer printout of the previous night's observation of an x-ray star in the constellation of Centaurus. Cen X-3, as this source was called, had first been observed by Frederick Seward's group at Lawrence Livermore Laboratory. Since then it had been observed by B. A. Cooke and Kenneth Pounds of the University of Leicester, and again by the Livermore group. As with Cygnus X-1, the findings of different groups and different observations by the same group had been conflicting. Cen X-3, then, was a good candidate for finding the type of short-term variability that had been discovered for Cygnus X-1. What Schreier, Tananbaum, and Giacconi found was something similar to Cygnus X-1, yet profoundly different.

The intensity of Cen X-3 was varying on a short time scale, as with Cygnus X-1. But in contrast to Cygnus X-1, where it had been impossible to say for sure whether the variations were periodic, the Cen X-3 periodicity almost leapt off the printouts. Approximately every five seconds, the intensity of the source would rise sharply to a maximum, then fall off slowly. Had they found another pulsar, of the type observed in the Crab Nebula? Several points suggested not. First of all, there was no evidence of a radio pulsar, or of an extended nebula around Cen X-3. Secondly, the pulse period was not nearly as steady as the Crab pulsar's. Finally, the pulse period was over a hundred times longer than that of the Crab pulsar. This meant that if Cen X-3 was a rotating neutron star, its rotational energy would be ten thousand times less than that of the Crab pulsar. Yet Cen X-3 was putting out approximately the same amount of energy as the Crab pulsar. At that rate the rotational energy of the neutron-star pulsar would suffice to power the source for only about one month. Even taking into account the possibility that Cen X-3 was much closer to earth than the Crab pulsar, so that its total power output might be a hundred times less than the Crab's, the rotational energy supply for Cen X-3 would still last only ten years. Clearly Cen X-3 was a pulsar of a different kind. It was also clear that many more observations of Cen X-3 would be needed in the months ahead. The observing plan was revised accordingly: Cen X-3 would be observed again in February and March.

The observation plan and the analysis and interpretation of the data were thrashed out in discussions among the scientists involved, and in more formal weekly meetings of the members of the AS&E scientific staff who were working on the project. These *Uhuru* meetings, as they were called, were held every Friday afternoon in Kellogg's office. In these often stormy sessions, constructive criticism was the order of the day; personalities sometimes clashed, but more often the clash was of an intellectual nature, as ideas, however wild, were floated with abandon and shot down remorselessly. The *Uhuru* meetings left most of the participants drained but also purged of any animosities they might have harbored as the result of a week of working together too closely. Although it often seemed quite the opposite immediately after a meeting, in the final analysis these meetings had the effect of building our confidence, both in our results and in one another.

The observations of Cen X-3 in February and March proved disappointing. The source was very weak and not pulsing. Still more data would be needed to unravel the complexities of Cen X-3. One week in May was set aside for *Uhuru* to observe Cen X-3 continuously. It was during this week, from May 4 to May 11, that the pieces began to fall into place for the puzzle of the x-ray stars.

With a data base of thousands of pulsations, it became possible to analyze the curiously shifting period to see if it fit any regular pattern. Schreier, working with Richard Levinson, Tananbaum, and Giacconi, soon established that the changes in the periodicity are themselves periodic. This is exactly what should happen if the x-ray pulsar is a member of a binary system. Unless the orbit is inclined at right angles to the observer, the x-ray star moves alternately toward and away from the observer as it circles its companion star. When the x-ray star is moving away from the observer, the pulse frequency decreases in accord with the Doppler effect; that is, the pulse period gets longer. As the x-ray star moves toward the observer, the pulses get closer together; that is, the period gets shorter. The pulses from Cen X-3 vary continuously in a manner that fits perfectly with the binary-star model.

The binary nature of Cen X-3 was soon confirmed by another discovery. Early in the day on May 6, the source disappeared. Then, about half a day later, it reappeared. Two days later, the source again disappeared and reappeared. By now the interpretation was obvious. The x-ray star was being eclipsed by its companion. Subsequently the optical counterpart of Cen X-3 was discovered by Wojciech Krzeminski of the Carnegie Institute of Washington. It is a massive blue star that shows optical evidence of a binary period of 2.1 days, the same as that of the x-ray star.

During the next few months another binary x-ray pulsar was discovered and extensively studied. This source, Hercules X-1, had a pulse period of 1.24 seconds and an orbital period of 1.7 days. The shortness of the pulse period effectively eliminates the possibility that the x-rays are produced by accretion onto a white dwarf, although detailed studies of the results from *Uhuru* and other satellites (*Uhuru's* successor, NASA's third Small Astronomy Satellite, *SAS-3;* Britain's *Ariel 3;* NASA's *OSO-8* and *Copernicus* satellites) were necessary to establish this conclusively.

Accreting neutron stars are in a way the opposite of radio pulsars. Matter is drawn onto the surface of the star instead of being expelled from it; this has the effect of speeding up the rotation of the star rather than slowing it down. This happens because the material spirals in with the same direction of rotation as the neutron star, so it adds to the star's rotational energy. Radio pulsars and binary x-ray pulsars are also opposite in that radio pulsars are seldom detected as periodically pulsing x-ray sources and binary x-ray pulsars are seldom detected as periodically pulsing radio sources. There are, however, a few binary x-ray stars that have been detected as weak radio sources. In these cases the radio emission, which is much weaker than that from a young supernova remnant such as the Crab Nebula, is believed to come from a large volume about the size of the binary system as a whole; it fluctuates on a scale of days or weeks

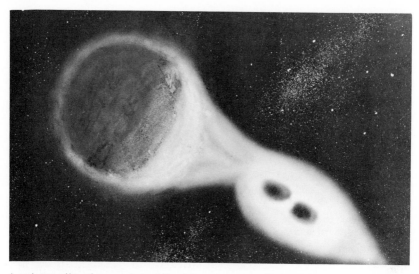

A schematic of a model of the binary x-ray pulsar Hercules X-1.
(Robert Plourde)

but does not pulse periodically on a scale of seconds. Any periodic radio emission that might exist near the surface of the neutron star is presumably quenched by the high density of infalling matter.

The only object firmly established to be both a radio and an x-ray pulsar is the Crab Nebula pulsar; this young neutron star is surrounded by a vast cloud of high-energy particles that is a bright source of radio, optical, and x-radiation. The binary x-rays stars are surrounded by no such clouds. This indicates that they are old neutron stars. Any cloud of high-energy particles generated by the supernova explosion that produced the neutron star has dissipated in the million or so years since they were formed.

A general scenario for the formation of a binary x-ray pulsar has been described by the Dutch astrophysicists Edward Van den Heuvel and John Heise, among others. Originally a double star system is formed, in which one of the stars, call it A, has, for example, a mass sixteen times that of the sun; the mass of the other star, call it B, is three times that of the sun. Over about six million years, star A will evolve to the giant phase. The gravitational force of star B will then begin to pull matter from star A. This will continue for about twenty thousand years, during which time star A will lose 75 percent of its mass to star B. Since B is not a collapsed star, the gravity at its surface is not unusually strong, and large fluxes of x-rays are not generated as the matter accretes onto its surface. After a few million more years, star A will explode, leaving behind a neutron star. When star B reaches the giant phase six or seven million years later, the gravitational field of the neutron star will pull matter off star B onto star A. Since A is now a neutron star, the gravity is enormous at its surface, and the infalling matter is heated to millions of degrees. The result is a strong source of x-rays that is observed as a binary x-ray pulsar.

In two important respects, binary x-ray pulsars and radio pulsars are similar. They are both neutron stars, and they both give off periodic pulses of radiation. In both cases, the magnetic field of the neutron star plays an essential role in producing the pulses. In the radio pulsar, the magnetic field provides the acceleration mechanism and a channel for the outflowing matter. For binary x-ray pulsars, the basic model was first worked out by Martin Rees and James Pringle at Cambridge University and by Rashid Sunyaev and his colleagues at Moscow University. The magnetic field acts as a funnel to guide infalling matter to a localized spot on or near the surface of the neutron star. It is in this funnel, or perhaps on the surface of the neutron star, that the x-rays are created. As the neutron star rotates, this hot spot comes into view once every rotation period, and a pulsing x-ray source is produced.

The first direct measurement of the strength of the magnetic field near the surface of a neutron star came from observations, using a detector carried in a balloon, of high-energy x-rays from the binary x-ray pulsar Hercules X-1. A research team led by Joachim Trumper at the Max Planck Institute detected a feature in the x-ray spectrum that could be identified as coming from electrons spiraling in a magnetic field. The wavelength of this feature indicated the strength of the magnetic field; it is several trillion times stronger than the magnetic field on the surface of the earth, and more than a billion times stronger than the strongest field observed on the surface of the sun.

The regular pulses produced by binary x-ray pulsars give astrophysicists an extremely useful tool for analyzing these fascinating systems. For example, careful study of the systematic increases and decreases of the pulse period makes it possible to deduce the motion of the pulsar. When this knowledge is combined with observations of the motion of the companion star and observations of the eclipses of the x-ray pulsar, the masses of the stars can be deduced.

The results obtained in this way are the only reliable measurements of the masses of neutron stars. John Bahcall of the Institute for Advanced Study and Yoram Avni of the Weizmann Institute of Science in Israel, together with a number of their colleagues, have refined this method through extensive calculations. In the most exhaustive study of this type, looking at seven star systems for which detailed observations are available, Saul Rappaport and Paul Joss of MIT have shown that the masses of the neutron stars fall between 1.2 and 1.4 times the mass of the sun. This is in good agreement with theoretical calculations of the expected mass of a neutron star, and the finding is a major confirmation of the theory of collapsed stars. While the observations of Cen X-3 were revealing the binary pulsar nature of that source, further observations of Cygnus X-1 were bearing fruit. Rocket flights by a group at Goddard Space Flight Center led by Stephen Holt and an MIT group led by Rappaport had found no evidence for a 73-millisecond period. Instead they found rapid, random fluctuations on a scale of 50 milliseconds. A reanalysis of old data by the NRL group yielded a similar result. Cygnus X-1 was clearly not an x-ray pulsar. Schreier and his AS&E colleagues made a detailed study of the large amount of data that had been obtained from *Uhuru.* They concluded that while the pulses were not strictly periodic, there were trains of pulses lasting tens of seconds with periods ranging from 0.3 seconds to tens of seconds.

On the flight on which they observed the time variability, the MIT group had used a collimator arrangement to obtain a more precise location for Cygnus X-1. This spurred radio astronomers to search for a radio counterpart. Robert Hjellming and Campbell Wade of the National Radio Astronomy Observatory and Luc Braes and George Miley of the Westerbork Observatory in the Netherlands independently identified a pointlike, randomly fluctuating radio source with Cygnus X-1. These powerful telescopes were able to locate the source within about one arc second, which is roughly equivalent to the angle subtended by a dime at a distance of one mile.

This precise location made it possible for optical astronomers to find the optical counterpart. A bright blue supergiant star called HDE 226868, for its number in the Henry Draper star catalog, was found very close to the position of the radio and x-ray sources. Astronomers Louise Webster and Paul Murdin of the Royal Greenwich Observatory suggested that this star was the companion of

A schematic of a model for Cygnus X-1, an x-ray source thought to be associated with a black hole. (Lois Cohen, Griffith Observatory; Courtesy TRW)

Cygnus X-1 and began a detailed study of its properties. They showed that HDE 226868 undergoes the telltale to-and-fro motion characteristic of stars orbiting in a binary system; it orbits an invisible companion star every 5.6 days. The mass of HDE 226868 was difficult to estimate exactly, but all indications were that it had a mass twenty times that of the sun, and that the unseen companion star had a mass about ten times that of the sun. Four weeks after Webster and Murdin published their results, C. T. Bolton of the University of Toronto published the results of an independent study of HDE 226868 that confirmed their conclusions.

The announcement of these results created a sensation in the astronomical community. According to the best theoretical estimates, a black hole accreting matter from a nearby companion star would be a strong source of x-radiation, and it should show evidence of a large mass (that is, greater than three times the mass of the sun, the maximum theoretically allowed mass of a neutron star) concentrated into a small volume. The x-ray source appeared to fit this description. It is part of a binary system in which a blue supergiant star is in a close orbit around an invisible companion star. The invisible companion star has a mass greater than about ten times the mass of the sun, and it is a strong x-ray source that shows rapid time variations in the intensity of its x-ray luminosity.

Because of the close correspondence between what was expected and what was observed, and because astrophysicists could think of no other object that could explain the observations, many astronomers cautiously concluded that Cygnus X-1 is indeed a black hole. There are four main points in the case for the identification of Cygnus X-1 as a black hole: (1) HDE 226868 is the optical counterpart of Cygnus X-1; (2) the mass of HDE 226868 is greater than twenty times the mass of the sun (this large mass guarantees, from considerations of the orbital dynamics of the system, that the invisible companion has a mass greater than the maximum permissible mass for a neutron star); (3) the x-ray source is compact; and (4) the only massive object we know of theoretically that is stable, compact, and capable of producing the observed x-ray emission is a black hole.

The identification of HDE 226868 as the companion star to Cygnus X-1 was based first of all on the coincidence of positions between the x-ray star and the blue star. Since the x-ray position had not been established with the precision of the optical and radio locations, more evidence was needed. One possibility was that the x-rays would show evidence of eclipses as Cygnus X-1 passed behind the blue star. In December 1971 and January 1972, *Uhuru* monitored Cygnus X-1 continuously for thirty-five days. No evidence was found of eclipses or any other variability associated with the blue star's orbital period of 5.6 days. Apparently the double star system is aligned so that we on earth are looking down on the plane of the orbit, so the x-ray source is not eclipsed. Later, F.K. Li and George Clark of MIT found evidence of absorption of x-rays when HDE 226868 was in front of Cygnus X-1; Keith Mason and colleagues of the University College of London found similar evidence when analyzing data from their x-ray detectors (which included three small x-ray telescopes) aboard the *Copernicus* satellite. It is thought that x-rays from Cygnus X-1 are absorbed by a stream of matter flowing from the blue star toward the black hole.

Tananbaum and his colleagues had found earlier evidence of the association of Cygnus X-1 with the blue star in a comparison of the radio and x-ray observations. They found that the x-ray intensity dropped dramatically between March 22 and March 31. It was during this time that the radio source associated with Cygnus X-1 first appeared. This correlated radio and x-ray behavior was strong evidence that the x-ray and radio sources were associated with the same object, Cygnus X-1. Since the position of the radio source was known to be coincident with the blue star to high precision, the association of Cygnus X-1 with the blue star HDE 226868 was secured. Much later, in 1978, the *Einstein* x-ray telescope located Cygnus X-1 with a precision of one arc second. As expected, its location was identical to that of HDE 226868.

The large mass of HDE 226868 was established by detailed optical studies of the star, and of nearby stars. Interstellar space, though almost empty, is not completely devoid of matter. Small dust particles in it redden starlight as it travels across the vast distances between stars, in much the same way as dust particles in the atmosphere redden sunlight at dusk. The amount of reddening of the light from a given star depends on the distance to the star; measuring the reddening of stars is a way to estimate their distances. In this way a group of astronomers at Lick Observatory, led by Joel Bregman, and a group at the University of California at Berkeley, led by Bruce Margon, showed that the distance to the Cygnus X-1 system is greater than seven thousand light-years. After estimating the distance, it was possible to estimate the total power produced by the blue star; this information indicated that its mass is greater than twenty times that of the sun. A study of the orbital motion of the blue star made it possible to estimate the mass of the companion star, Cygnus X-1: this mass is greater than six times the mass of the sun, and more than twice as great as the maximum possible mass of a neutron star.

The rapid, random changes in the intensity of the x-ray emission establish the compactness of the source. The flaring region cannot be so large that the time it would take for a disturbance to travel across the region is longer than the observed time scale for the flaring. Otherwise the entire region would not change its brightness coherently, or all at once, and detectable variation would occur. Since no disturbance can travel faster than the speed of light, it follows that if a flare occurs in a second, the size of the flaring source must be less than a light-second (the distance light travels through space in a second: 300,000 kilometers), and if a flare occurs in a tenth of a second, the source must be smaller than a tenth of a light-second. The observed 50-millisecond fluctuations of x-ray intensity in Cygnus X-1 mean that the source must have a diameter less than or equal to that of a white dwarf. Thus either the star responsible for the x-ray emission is a collapsed star such as a white dwarf, a neutron star, or a black hole, or else the radiation must come from a small active region on a larger noncollapsed star. But there is no evidence for a second noncollapsed star in the system, and no viable mechanism has been proposed for producing the enormous x-ray luminosity of Cygnus X-1 from an active region on a normal star or from the region between two stars in a binary system. The infall of matter onto a collapsed star, however, provides a natural explanation of the x-ray emission, which is about ten thousand times as powerful as the total power output of the sun. Therefore, we are left with a collapsed star as the only plausible explanation for the intensity and variability of the emission of x-rays, and the only known type of collapsed star that can have a mass greater than three times that of the sun is a black hole.

By the end of 1972, the consensus among astronomers was that the riddle of the x-ray stars had been solved. They are binary star systems in which matter is streaming from a normal star onto a nearby collapsed star. In most cases the collapsed star is a neutron star, but in at least one and probably two cases, the collapsed star is most likely a black hole. The two-year period after the launch of *Uhuru* was one of those rare exhilarating times in science when, as the result of a great idea, or, as in this case, a sudden increase in observational capability, the fog lifts and we can see clearly another portion of the cosmic landscape.

Adding to this excitement was the discovery of very strong evidence for a new type of object, the black hole. Black holes represent something absolute in a world in which very little is absolute; they have captured the imaginations of scientists and laymen alike, and have quickly found their way into everyday language and popular literature. How common are black holes? The search for black holes among the x-ray stars has led to another very likely possibility. Recently, a team of Canadian and American astronomers have used optical observations and results from two of NASA's High-Energy Astronomical Observatories to show that the strong, highly variable x-ray source LMC X-3, in the Large Magellanic Cloud (a nearby galaxy), is associated with a dark star with a mass greater than nine times the mass of the sun. Since this mass far exceeds the limit of three solar masses for neutron stars, this object must be considered a strong candidate for a black hole.

We suspect that other, much more massive black holes exist in the centers of galaxies. A black hole might form in the center of a galaxy as a result of collisions of hundreds of thousands of massive stars in a dense swarm of stars, or perhaps from the collapse of a single, supermassive star. Consideration of these possibilities takes us beyond individual stars to the realm of the galaxies.

8

The X-Ray Sky

The word *galaxy* comes from the Greek *galaxias*, meaning "milky circle" or, more familiarly, "milky way." Although the Milky Way has been known and described, at least poetically, for more than two thousand years, it was not until the twentieth century that astronomers could be confident of their scientific description of it, or could say whether it was unique or commonplace in the universe.

One of Galileo's first accomplishments with the telescope was to show that the Milky Way is not a continuous band of light but a multitude of stars that appear to the naked eye to blend into a milky band because of their great distance. By the end of the eighteenth century, the English astronomer William Herschel had concluded, correctly, that our solar system is part of a large disk-shaped system of stars, and that the bright band of the Milky Way is the effect of viewing this system, or galaxy, as it came to be called, along the plane of the disk. In the twentieth century, astronomers began to understand the dimensions of the galaxy and the location of the solar system within it. In 1917 Harlow Shapley of the Harvard College Observatory showed that the sun is located many thousands of light-years from the center of the galaxy. He estimated the distance of the sun from the center to be 50,000 light-years and the diameter of the galactic system to be 250,000 light-years. In 1930 Robert Trumpler showed that the dimming of starlight by interstellar dust had caused Shapley to overestimate the size of the galaxy. Shapley had assumed that the dimness of the stars and star clusters he was studying was solely an effect of distance. In dust-free space, a star that is four times dimmer than a similar star must be twice as distant. But if dust causes part of the dimming, the dimmer star may not be twice as far away. Taking this effect into account, the sun is about 30,000 light-years from the center of the galaxy, whose disk has a diameter of 90,000 light-years.

The overall shape of our galaxy is a disk with a central bulge. The stars in the disk orbit around the center of the galaxy; the solar system takes about 220 million years to complete one orbit. The galaxy contains a few hundred billion stars. The young, bright stars are concentrated in the disk, as are the interstellar dust and gas. The older stars, which have low mass and are relatively dim, are located in the central bulge and in a more or less spherical halo with a diameter about twice that of the disk.

Another major accomplishment of early-twentieth-century astronomy was the discovery that ours is not the only galaxy in the universe. In the 1920s Edwin Hubble of the California Institute of Technology took advantage of a significant advance in telescope technology, the inauguration of the 100-inch telescope on Mount Wilson, to show that the Milky Way galaxy is not unique. In every direction, and as far as the most powerful telescopes can see, there are galaxies of different sizes and shapes.

A galaxy is a swarm of stars, gas, and dust bound together by their mutual gravitational attraction. The character of a galaxy depends on how many and what type of stars are in the swarm, how much gas and dust are present, and how rapidly the galaxy is rotating, among other things. Our galaxy has a good number of massive young bright stars and a relatively large (10 percent by mass) amount of gas and dust, and it is rotating fairly rapidly, which is presumably the cause of its flat spiral shape. Other galaxies have a roughly spherical or elliptical shape. These so-called elliptical galaxies contain mostly old, low-mass stars, few massive young bright stars, and little dust and gas, and they rotate slowly if at all. Elliptical galaxies range in size from dwarf ellipticals, which have fewer than one percent as many stars as the Milky Way galaxy, to giant ellipticals, which can have many times more stars than our galaxy. In between the dwarf ellipticals, the spirals, and the giant ellipticals is a continuous range of sizes and shapes.

Our galaxy's nearest neighbors are the Magellanic Clouds, two small irregularly shaped galaxies visible in the southern hemisphere at distances of 160,000 and 200,000 light-years. The nearest galaxy comparable to ours is the Andromeda galaxy, which is about two million light-years away. This giant spiral galaxy has roughly the same shape and the same number and types of stars as the Milky Way.

Galaxies, like stars, can swarm together to form larger systems called groups and clusters, and clusters of galaxies apparently group together into superclusters. Our galaxy, along with the Andromeda galaxy and about twenty smaller galaxies, form what is called the Local Group. Most of the galaxies in the universe are found either in isolation or in small groups. Clusters of galaxies are of great interest to astronomers, however, because they provide opportunities for studying the interactions among galaxies; also, because they can be seen at great distances, they can be used as tools for studying the large-scale structure of the universe.

The nearest rich cluster of galaxies is about sixty-million light-years away, in the direction of the constellation of Virgo. This system contains several thousand galaxies; the most common types in the cluster are large spirals and smaller elliptical galaxies, but the most spectacular are several giant elliptical galaxies, especially one we know as M87. M87 is the eighty-seventh object listed by Charles Messier, a French astronomer who was mainly interested in finding comets, and who cataloged all bright diffuse objects that he thought might be mistaken for comets. Ironically, many of these objects Messier wished to avoid, such as the Crab Nebula (M1), the Andromeda galaxy (M31), and M87, turned out to be of far more enduring interest than any of the comets he discovered.

M87 is located in the center of the Virgo cluster of galaxies. It is estimated to contain as many as a trillion stars. The center of the galaxy is completely overexposed on normal astronomical photographs because of the high density of stars there, but on short exposures a peculiar jet-like feature is visible. Since the 1950s, this jet has been known to be a strong source of radio and optical radiation, and by implication, of high-energy particles. The jet comes out of the central region, or nucleus, of the galaxy; it was apparently ejected from the nucleus a million or so years ago in a burst of explosive activity that must still be occurring.

Optical and radio studies indicate that some level of explosive activity occurs in almost every galactic nucleus. Galaxies in which the explosive activity is especially pronounced are called active galaxies. Quasars are thought to be the most extreme examples of active galaxies. Quasars radiate energy at a rate of ten trillion suns, or about a thousand times more than our entire galaxy, from a region approximately the size of our solar system.

Looking at the sky through an optical telescope, we see glowing clouds of gas and stars of various colors and intensities nearby; they merge to form the Milky Way as we view them from larger distances. Beyond the Milky Way we encounter galaxies, clusters of galaxies, and quasars.

By the end of the *Uhuru* mission, a rough picture of the x-ray sky was available. Under the direction of Steve Murray of AS&E, a catalog of more than a hundred x-ray sources was compiled. From this catalog several features were apparent. First of all, at an equal level of received power per unit area at optical wavelengths, some two million stars and eight thousand galaxies are visible. In x-rays we are seeing the outline or skeleton of the galaxy. The situation is not unlike a comparison of x-ray and optical photographs of humans: different aspects of the subjects are revealed by the photographs at different wavelengths. Photographs of the sky at visible wavelengths reveal, in essence, the "middle age" of stars. They reveal stars that shine as a result of the burning of nuclear fuel deep

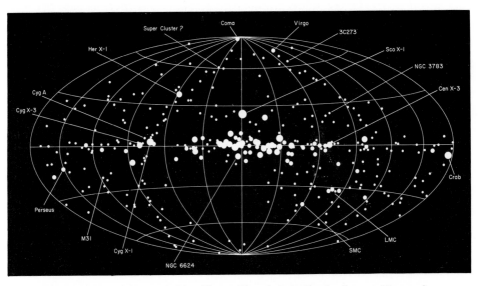

The x-ray sky as revealed by *Uhuru*. The dots indicate the positions of the x-ray sources in the Fourth *Uhuru* Catalog of sources. The map is in galactic coordinates, so the central plane of the galaxy is aligned with the equator and the center of the galaxy is in the center. (H. Tananbaum, Harvard-Smithsonian Center for Astrophysics)

in their interiors. X-ray observations, by contrast, help us to understand the last stages in the evolution of stars; a time when nuclear fuels have been exhausted and a star has collapsed, sometimes catastrophically, to become a white dwarf, a neutron star, or a black hole. A massive star first becomes a strong source of x-rays shortly after it collapses; in the case of neutron stars and possibly black holes, the glowing remnant of a supernova is formed. For some remnants a rapidly rotating neutron star supplies a stream of energetic particles that produces an extended cloud emitting radio, optical and x-radiation. The Crab Nebula is an example of an x-ray source of this type. Supernovas also produce x-rays as the matter ejected in the explosion interacts with the surrounding interstellar gas. This interaction generates a shock wave; the hot gas behind this shock wave glows in x-rays for several thousand years.

After about a hundred thousand years, all traces of x-ray emission from a supernova explosion will have vanished. Then a strong source of x-rays can be produced only by the accretion of an appreciable amount of matter onto a collapsed star. X-ray stars are rare, not because collapsed stars are rare but because the production of x-rays requires a source of material to accrete, which in turn requires the presence of a nearby companion star, a comparatively rare situation. Most of the strong x-ray sources in the galaxy are of this type.

Only three normal galaxies showed up on the *Uhuru* sky map: the Magellanic Cloud galaxies and the Andromeda galaxy. This was expected because of the rarity of x-ray stars, which account for most of the x-radiation from normal galaxies. In active galaxies, by contrast, the nucleus of the galaxy is unusually bright in both visible and x-radiation. The source of this strong visible and x-radiation cannot be understood in terms of normal stars and x-ray stars. A different process, one that produces intense radiation over a wide range of wavelengths, must be at work. In 1966 E. T. Byram, Talbott Chubb and Herbert Friedman of NRL detected x-radiation from the giant elliptical galaxies M87 and NGC 1275. Subsequently Stuart Bowyer and his colleagues at the University of California at Berkeley detected the quasar 3C273. It was not surprising, then, that several active galaxies and one quasar showed up on the *Uhuru* sky map.

One major surprise was the discovery of intense x-ray emission from clusters of galaxies. Largely as a result of work by Herbert Gursky, Edwin Kellogg, William Forman, and Alfonso Cavaliere, twenty sources detected by *Uhuru* were identified with clusters of galaxies. The radiation from these sources came from an extended region more than a million light-years across and, judging from the spectrum of the radiation, was apparently a property of the cluster as a whole rather than radiation from the individual galaxies. In 1966 a group at Goddard Space Flight Center reported observations from a rocket experiment that indicated x-ray emission from a rich cluster of galaxies in the Coma constellation. This led astrophysicist James Felten and his colleagues at the University of California at San Diego to suggest that the x-radiation is produced by a large reservoir of hot gas trapped in the cluster and heated by the large random velocities of the galaxies. Subsequent attempts by other groups to confirm the source failed, however, and the idea that clusters of galaxies could be x-ray sources was set aside until the *Uhuru* discoveries.

After the *Uhuru* results were published, James Gott and James Gunn of Princeton University proposed that the radiation from clusters of galaxies is caused by intergalactic gas that has been captured in the gravitational field of the cluster and heated by compression as it falls into the cluster. Other scientists noted that clusters of galaxies are sources of radio waves as well as of x-rays. The radio and x-ray emissions occur in overlapping regions of space. Some astronomers concluded from this that the two phenomena are related; perhaps the x-rays are produced by the interaction of the high-energy electrons responsible for the radio waves with the microwave background photons that fill all space. This interaction process is called Compton scattering, and these models are called Compton scattering models.

The data on clusters of galaxies indicated that richer clusters, that is, the ones with more galaxies, are more likely to be x-ray sources. These clusters, having more galaxies than poorer clusters, have more mass concentrated in roughly the same volume. Therefore their gravitational field is stronger. Since the average velocity of the galaxies within a cluster depends on the strength of the gravitational field, stronger gravity means larger velocities on the average. Since richer clusters appear to be stronger x-ray sources, Alan Solinger and Wallace Tucker of AS&E suggested that there might be a correlation between the power radiated in x-rays and the average velocity of the galaxies in the cluster. They found evidence in the *Uhuru* data for such a correlation and argued that the details of the correlation supported hot-gas models for the x-ray emission from clusters of galaxies.

The *Uhuru* data were not sufficient to decide between the hot-gas and Compton scattering models. But one class of hot-gas models made a prediction that could be tested by future experiments. If the gas producing the x-ray emission comes from the stars in the galaxies in the cluster, then it should be mostly hydrogen with small traces of heavier elements such as silicon, sulfur, and iron; the x-ray spectrum from the hot gas in the cluster should show evidence of one or more of these elements.

In the seven years following the launch of *Uhuru*, seven satellites with x-ray detectors were flown. These included a small x-ray telescope with a mirror about the size of a half-dollar on NASA's Orbiting Astronomical Observatory *Copernicus* satellite, experiments aboard two of NASA's Orbiting Solar Observatories (OSO's) and the Defense Department's *Vela 5-A* satellite, the Astronomical Netherlands Satellite *(ANS)*, the British *Ariel 5* satellite, and one of *Uhuru*'s successors in the Small Astronomy Satellite series, *SAS-3*. In addition, a number of important balloon and rocket experiments were performed. The greater sensitivity and extended wavelength

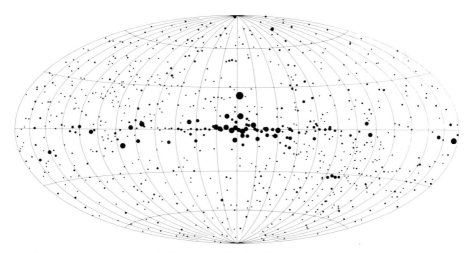

The x-ray sky as revealed by the HEAO A-1 x-ray detector aboard NASA's first High-Energy Astrophysical Observatory *(HEAO-1)*. (K. Wood, Naval Research Laboratory)

coverage of these experiments enhanced and clarified our knowledge of the x-ray sky.

More new binary x-ray pulsars were discovered by the *Ariel 5,
SAS-3, OSO-8,* and a NRL rocket. Observations of these sources and
the already-known binary pulsars by these satellites and by balloons
established beyond doubt that the binary x-ray pulsars are accreting
neutron stars.

During this time Jonathan Grindlay of the Harvard-Smithsonian
Center for Astrophysics discovered, while analyzing data from the
ANS satellite, an x-ray star that was giving off sudden bursts of
x-rays. Within months, *Vela-5A, SAS-3, Ariel 5,* and *OSO-8* had
discovered similar bursts from about three dozen other x-ray stars.
These "x-ray bursters" give off as much energy in x-rays in a few
seconds as the total energy output of the sun in two weeks. The
bursts follow no well-defined pattern, but recur sporadically at intervals ranging from several hours to a few days. Two possible
models were put forward to explain the bursters. On the one hand, a
sudden increase in the amount of matter falling onto the surface of
the neutron star could produce a sudden increase in x-ray emission.
On the other hand, accreted matter that had accumulated on the
surface of the neutron star could, under the right conditions, undergo a thermonuclear explosion, producing a burst of x-rays.

A combination of theoretical work and detailed analysis of the
satellite data, especially by Walter Lewin and Paul Joss and their
colleagues at MIT using *SAS-3* data and by Minoru Oda of the University of Tokyo and his colleagues using data from the Japanese
Hakucho satellite, has shown that most if not all the x-ray bursts are
produced by thermonuclear explosions on the surface of neutron
stars. Observations indicate that x-ray bursters do not pulse and
that binary x-ray pulsars do not produce bursts. The key to whether
a neutron star accreting matter from a nearby companion will be a
burster or a pulsar appears to be the strength of the magnetic field
near the surface of the neutron star. For relatively strong magnetic
fields, the matter is channeled onto a small region of the star. This,
combined with the star's rotation, causes x-ray pulses; it also produces a higher density and temperature of the accreting matter,
which causes the hydrogen and helium to undergo steady thermonuclear fusion; this prevents the accumulation of explosive
amounts of hydrogen and helium, so that bursts do not occur. For a
relatively weak magnetic field, the magnetic funneling of matter is
ineffective, so no pulses are produced; the matter settles onto the

surface of the star more uniformly and at lower densities; steady thermonuclear fusion does not occur, and concentrations of hydrogen and helium build up on the surface. When this stockpile exceeds a certain critical mass, a thermonuclear explosion occurs, producing a burst of x-rays.

It was also during this period that some of the objects familiar to traditional optical astronomers began to show up in the x-ray sky. In March 1973 an MIT rocket experiment by Saul Rappaport and his colleagues detected x-rays from SS Cygni, a white dwarf in a close binary system. *SAS*-3 detected x-rays from other white dwarfs, as did a rocket experiment by Bruce Margon and his colleagues.

The first detection of x-rays from a noncollapsed star was made by Richard Catura, Loren Acton, and Hugh Johnson of Lockheed in a rocket experiment in April 1974; they observed soft x-ray emission from Capella, which is visually the sixth brightest star in our sky. After *SAS*-3 observed x-rays from a few other normal, that is, noncollapsed stars, it became clear that as more sensitive detectors were used, many normal stars would be found to emit x-rays, and x-ray astronomy would make important contributions to the mainstream of stellar astronomy.

Going beyond the galaxy, *Ariel* 5 and *SAS*-3 demonstrated that many active galactic nuclei are strong emitters of x-rays. *SAS*-3 also discovered two more quasars. *Ariel* 5 made another significant discovery. In analyzing data from *Ariel* 5, R. J. Mitchell and his colleagues at Mullard Space Science Laboratory of the University College of London detected a feature in the spectrum of the Perseus cluster of galaxies that they identified with emission from highly ionized iron atoms — iron atoms that have been stripped of 24 of the 26 electrons that normally orbit the nucleus. Looking at data from the *OSO*-8 satellite, Peter Serlemitsos and his colleagues at Goddard Space Flight Center confirmed these findings and detected similar features from the Virgo and Coma clusters. Subsequently Mitchell and Leonard Culhane of the Mullard Laboratory group found such emission from highly ionized iron atoms in yet another cluster in the constellation of Centaurus. These results demonstrated conclusively that the x-ray emission from these clusters of galaxies, and by implication from all clusters of galaxies, comes from hot gas. The percentage of iron implied by the observations is roughly the same as the percentage of iron found in the sun and in the interstellar gas. Astrophysical theories of the origin of the elements indicate that all the elements heavier than helium have been manufactured inside stars. Supernova explosions then eject this material from the stars, thereby enriching the gas in and around galaxies with heavy elements. The space between clusters of galaxies has presumably not been enriched and should not be contaminated with heavy elements such as iron. The discovery of appreciable concentrations of iron in clusters of galaxies indicated that most of the hot gas in the clusters came from galaxies and not from intergalactic space.

The era of small satellites in x-ray astronomy came to an end on August 12, 1977, with the launch of NASA's first High-Energy Astronomy Observatory satellite, *HEAO-1*. This 3½-ton satellite was in essence a super *Uhuru* (in fact, when first proposed it was called a Super Explorer). It carried instruments designed to map the x-ray sky and to study individual sources of x-rays in detail. The instruments were able to locate x-ray sources and to study their variation in time and the distribution of their x-radiation with wavelength over a broad range of wavelengths. Principal investigators were Herbert Friedman of NRL, Elihu Boldt of Goddard, Gordon Garmire of Cal Tech, Herbert Gursky and Daniel Schwartz of Harvard-Smithsonian, Hale Bradt and Walter Lewin of MIT, and Lawrence Peterson of the University of California at San Diego.

The history of the HEAO program is an interesting example of how large science projects become a reality. HEAO was the product of more than a decade of dreaming, planning, political infighting, and hard work by a large number of scientists and engineers. It began to take shape in the early 1960s, primarily as the result of the dreams of three men. Frank McDonald of Goddard wanted a large astronomical satellite to carry a heavy cosmic-ray experiment, Herbert Friedman wanted a large satellite for a large system of proportional-counter x-ray detectors, and Riccardo Giacconi wanted one for a large x-ray telescope.

McDonald, Friedman, and Giacconi discussed the idea with other scientists, and in 1965 they received an endorsement from the influential Space Science Board. Meanwhile McDonald found important allies at NASA headquarters: Jesse Mitchell, director of the Geophysics and Astronomy Division, and Richard Halpern, who worked for Mitchell. In 1967 the HEAO concept received an essential endorsement from NASA's Astronomy Missions Board, a board of ten respected and knowledgeable members of the astronomical community. NASA appropriated funds for a feasibility study of a series of four large HEAO satellites. Halpern became program manager. Goddard Space Flight Center seemed the logical NASA center to handle the project, but Goddard was at the time heavily involved with other projects, so it refused. McDonald and Alois Schardt of NASA headquarters then approached the Marshall Space Flight Center in Huntsville, Alabama.

Under the direction of Werner von Braun, Marshall had risen to prominence among NASA centers as a rocket-building center during the big push of the Apollo program to put a man on the moon. In the late sixties, the peak of the Apollo effort was over as far as Marshall was concerned, so von Braun and his associates, including Ernst Stuhlinger, who was chief scientist at Marshall, were receptive to McDonald and Schardt's overtures. They agreed to support the HEAO project and appointed Carroll Daily of Marshall to help engineer the project and to help McDonald, Halpern, and Mitchell firm up support within NASA and in the Congress.

In 1970 Congress approved the HEAO program. Albert Opp of NASA headquarters was named program scientist, and Fred Speer was named program manager at Marshall. TRW in Los Angeles was selected as the industrial contractor to build the spacecrafts. The program moved ahead until January 2, 1973, when Halpern and others associated with the program were suddenly notified of its cancellation. The Nixon administration had cut NASA's budget in 1974; at the same time the Viking program to land a spacecraft on Mars was incurring large cost overruns. To get money to keep Viking alive, some program had to be cut, and the only one with enough money was HEAO. Accordingly, HEAO was canceled.

Mitchell and Halpern immediately set to work to have the cancellation reversed. First, with the help of engineers at TRW and Marshall, they put together a new, less expensive version of HEAO. These included Friedman's large-area proportional counters and three other cosmic x-ray experiments, but McDonald, one of the fathers of the HEAO concept, lost out. His cosmic-ray experiment was simply too heavy. He did, however, continue to serve as project scientist for the first HEAO satellite. The scientific community weighed in with protests emphasizing the importance of high-energy astronomy. Mitchell took his case to NASA's higher management, and he won. The cancellation was changed to an 18-month suspension. In July 1974 major funding began again, but at half the previous level. A little over three years later, shortly after midnight on August 12, 1977, an Atlas-Centaur rocket carrying the first of three HEAO payloads lifted off the launch pad. Twenty-three minutes later, more than a decade after it was first proposed, *HEAO-1* was in orbit.

HEAO-1 remained operational until January 9, 1979. During that time it made a greatly improved map of the x-ray sky. More accreting white dwarfs and x-ray-emitting normal stars were discovered and studied, as well as a spectacular feature in the constellation of Cygnus. This feature, called the Cygnus superbubble, was discovered by Webster Cash of the University of Colorado and his colleagues. The superbubble is a cloud of gas that stretches across more than a thousand light-years; it contains the gaseous mass of several hundred thousand suns. This feature was apparently produced by a chain reaction of twenty or more supernova explosions. In such a reaction, a massive star explodes in a cloud of cool gas, sending out shock waves that compress the cloud, triggering the formation of more massive stars that explode after a few million years, triggering the formation of still more stars, and so on, producing an expanding superbubble of hot gas.

HEAO-1 was able to measure the x-ray spectra of active galaxies, quasars, and clusters of galaxies over a wide range of energies, and it added many more of these objects to the x-ray sky. As an expanded, super *Uhuru*, it fulfilled its mission of surveying the entire sky and culminated the pretelescopic era of x-ray astronomy. Although *HEAO-1* was fifty times larger than *Uhuru*, its largest detector was only seven times more sensitive than *Uhuru*'s (the signal-to-noise ratio improves only with the square root of the area of this type of detector). Any further improvement in the sensitivity of x-ray detectors would have to be achieved by a different approach.

9

A Telescope for X-Rays

The introduction of new technology into any area, whether it is farming, steel milling, automobile manufacturing, warfare, or a particular field of science, is always a complex and difficult process that cannot occur until the time is right. The large x-ray telescope observatory was no exception. It took time to develop the necessary technology. It took time to convince the scientific community and the funding agencies of the need for such a mission. And it required a continuous struggle to build the observatory in the face of a multitude of technical and budgetary problems.

As early as 1960, Giacconi and Rossi published a description of a design for an x-ray telescope and discussed the enormous advantages of an x-ray telescope over conventional x-ray detectors. Under NASA sponsorship, Giacconi and Norman Harmon at AS&E began to build experimental x-ray telescopes. By July 1961 they had built an x-ray collector with an area of about one square centimeter, or about half the area of a dime. Six months later they had an x-ray telescope with roughly the same effective collecting area.

X-rays can be focused by reflection only if they strike the reflecting surface at grazing angles of less than a degree or so. Therefore the reflecting surface of an x-ray telescope cannot be perpendicular to the incoming beam, as in optical telescopes, but must be nearly parallel to the beam. The reflecting surfaces of an x-ray telescope look more like cylindrical tubes than like the dishes of radio and optical telescopes. The internal surfaces follow subtle parabolic and hyperbolic curves. The first x-ray telescopes constructed at AS&E were built by machining and polishing the interior surfaces of aluminum tubes. The surfaces were then coated with evaporated gold to provide high reflectivity. These telescopes were far from the theoretical ideal, but the polishing and grinding tools that had been used were relatively unrefined. Under these conditions, the AS&E group was elated by laboratory demonstrations that the telescopes focused x-rays, however imperfectly.

The next step was to improve upon these prototypes and build an x-ray telescope to send into space. But this required more money from NASA for research into new methods for preparing cylinders with smooth parabolic and hyperbolic inner surfaces. An analogy gives some idea of the smoothness required in a first-class x-ray telescope: if the earth were proportionately as free from bumps as a finely polished x-ray telescope, the highest mountain would be only six inches high. Such precision was already available in preparing mirrors for optical telescopes, and it could in all probability be obtained for the more difficult configurations of x-ray telescopes if money for research and development could be found. But the funds would be available only if NASA approved an x-ray telescope for a flight mission, and NASA was unlikely to approve an untested concept for a flight mission.

This impasse was broken with the help of John Lindsay of Goddard Space Flight Center. Lindsay had been a member of Herbert Friedman's solar x-ray astronomy group at NRL, and had moved to Goddard shortly after the establishment of NASA. At Goddard he had been a strong and effective advocate for solar x-ray astronomy. He was instrumental in starting the program of the Orbiting Solar Observatory series of satellites. He also served as technical monitor for NASA on the contract that funded the early AS&E research and development of x-ray telescopes. Lindsay was enthusiastic about the potential of x-ray telescopes for solar x-ray astronomy, and because of his position as an insider in the NASA establishment he was able to play a key role, along with Harold Glaser of the Solar Physics Branch of NASA headquarters, in persuading NASA to pursue the development of x-ray telescopes for use in solar x-ray astronomy.

Although the AS&E group's primary interest was in extrasolar x-ray astronomy, they were convinced that the future of x-ray astronomy lay in the development of x-ray telescopes. Accordingly, Giacconi initiated a program in solar x-ray astronomy at AS&E, with an emphasis on developing an x-ray telescope to study the x-ray emission from active regions and flares on the solar surface. A collaboration began between Lindsay and others at Goddard and

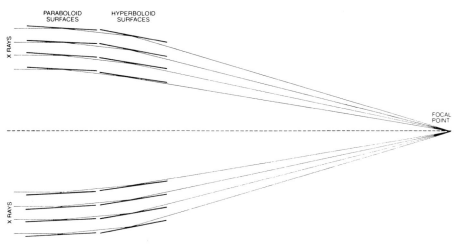

A schematic of a grazing-incidence telescope.

Giacconi and others at AS&E. On a rocket flight on October 15, 1963, and another on March 17, 1965, the group obtained the first x-ray pictures of the sun using a grazing-incidence telescope. These telescopes, made of electroformed nickel, had a diameter about equal to that of a tennis-ball can and a focal length of approximately 80 centimeters.

Lindsay and Giacconi planned a much more ambitious collaboration to build an x-ray telescope with a diameter of 26 centimeters for the Advanced Orbiting Solar Observatory. This telescope would have an angular resolution of 5 arc seconds, which would compare favorably with the 1-arc-second resolution of the large optical telescopes. In 1965 NASA canceled the Advanced Orbiting Solar Observatory, but the outlook for a large x-ray telescope remained good, as NASA began planning to incorporate it into a large manned spacecraft. Then in September 1965 John Lindsay died suddenly of a heart attack. The collaboration between AS&E and Goddard was dissolved. In the years to come, Lindsay's pioneering efforts would bear fruit in the *Skylab* mission, in which two x-ray telescopes, one built by an AS&E group, the other by a Goddard group, would send back spectacular x-ray images of the sun.

Despite continuing progress in the use of x-ray telescopes to study the sun, the adoption of x-ray telescopes to study x-ray sources outside the solar system proceeded at a snail's pace. NASA rejected proposals to develop small x-ray telescopes for use on rockets and larger telescopes for use on satellites. Two primary reasons for NASA's reluctance to adopt a new technology can be identified.

First of all, the x-ray stars were unusually bright, so that conventional techniques using mechanically collimated proportional counters were yielding an abundance of scientific returns. Simply using these techniques on satellites such as *Uhuru* yielded considerable gains in sensitivity. Thus NASA authorities saw no urgent need to develop x-ray telescopes.

Secondly, it was still not clear that x-ray telescopes would work on sources outside the solar system. For an intense source of x-rays such as the sun, the loss of detecting power caused by microscopic imperfections in the surface of a telescope is unimportant. Because of the great distances they travel, the fluxes of x-rays from outside the solar system are millions of times less than those from the sun; under these conditions, the loss of power caused by x-rays hitting imperfections in the surface and scattering can be crucial. Highly polished surfaces comparable to those of the best optical telescopes had to be achieved. It would also be desirable to nest the cylinders, one inside the other, to achieve the maximum possible reflecting area; this involved the technically delicate problem of co-aligning the nested mirrors. Finally, for use on an unmanned satellite, film could not be used as the detecting medium; instead it would be necessary to develop x-ray-imaging cameras capable of high resolution that could detect one x-ray photon at a time and record this information electronically.

From NASA's point of view, all this meant that the time was not yet right to plunge into an expensive program to develop an x-ray telescope sophisticated enough to do x-ray astronomy outside the solar system. From the point of view of Giacconi and his colleagues at AS&E, it meant that continued work on high-quality solar x-ray telescopes would not necessarily bring them closer to their goal of a high-quality extrasolar x-ray telescope. Yet funds were available for solar rather than extrasolar x-ray telescopes, so the AS&E group decided to make the best of the situation. In developing a solar x-ray telescope for the Apollo telescope mount on *Skylab*, they incorporated certain sophisticated features, such as a nested configuration of two mirrors. And they began to construct x-ray television cameras, although these were not strictly necessary for the mission, since film retrievable by astronauts was used. Thus, they were able to keep the technology of extrasolar x-ray telescopes moving forward.

NASA was not alone in its reluctance to embark on the development of a large x-ray telescope. The scientific community was also lukewarm to the idea. Some advisory committee members stressed the need to accomplish all-sky surveys with conventional detectors first to see if such a telescope was necessary. Others expressed similar views — perhaps x-ray astronomy would not turn out to be that interesting, so why rush? Still others stated flatly that theoretical considerations indicated that x-ray sources were mainly diffuse, extended regions, so that there was no need for high-angular-resolution x-ray telescopes. Fortunately, the real world turned out to be far richer than this particular "world on paper."

In spite of this resistance, Giacconi continued to promote the concept of x-ray telescopes in any forum he could find. In Woods Hole, in the summer of 1965, the x-ray and gamma-ray panel of the Space Science Board of the National Academy of Sciences met to study plans for future x-ray and gamma-ray experiments. Among those on the panel were Frank McDonald of Goddard and many of the pioneers of gamma-ray and x-ray astronomy: Rossi and George Clark of MIT, Philip Fisher of Lockheed, William Kraushaar of the University of Wisconsin, Robert Novick of Columbia University, Giacconi, and Friedman, who was chairman of the panel. Giacconi used the occasion to generate a debate about the relative merits of sensitive conventional detectors and x-ray telescopes. He pointed out that conventional detectors have natural limitations that impose severe restrictions on sensitivity. These include the background noise, which is practically negligible in an x-ray telescope, and the problem of distinguishing sources. This is analogous to the difficulty a person with poor eyesight has in distinguishing the many words that appear on a printed page. With adequate lenses or, by analogy, with a telescope of adequate resolution, the problem can be greatly reduced or effectively eliminated.

Out of this debate, a compromise emerged. The panel agreed that x-ray telescopes based on grazing-incidence reflection "appear to be the only tools capable of many of the refined observations that will be needed beyond the early exploratory stage," and they recommended that "a program of x-ray astronomy using total reflection telescopes should be started at the earliest possible time and pushed vigorously to exploit its ultimate capability." They also recommended that a survey mission carrying a large-area x-ray detector of the type promoted by Herbert Friedman be flown before the x-ray telescope, and they endorsed a large cosmic-ray experiment of the type envisaged by Frank McDonald. As discussed earlier, this report laid the groundwork for NASA's High Energy Astronomical Observatory (HEAO) program, which would involve a series of large but inexpensive spacecraft that could accommodate the experiments.

While McDonald went to work inside NASA promoting the general HEAO concept, Giacconi continued to push for the more immediate goal of a smaller x-ray telescope on the somewhat smaller Orbiting Astronomical Observatory (OAO) spacecraft. This would be followed, he proposed, by a larger, more ambitious x-ray observatory, which would be built and operated by a consortium of scientists from various institutions. The proposers expressed a willingness to let other astronomers, not in the consortium, use the facility as guest observers. This was essentially the way the major optical and radio observatories operated, but it was a major departure for space science. X-ray astronomy in particular had been distinguished by competition rather than cooperation. NASA rejected the idea of an OAO-type mission involving a single institution, but it expressed interest in the larger facility. Meetings were held, and the idea was discussed almost to the point of extinction, it seemed. Yet slowly but surely momentum for a large x-ray telescope began to build. This movement was assisted in no small measure by McDonald's efforts on behalf of HEAO. If NASA approved a series of large spacecraft missions, they would have to put something on them, and a large x-ray telescope was a distinct possibility. In February 1968, NASA issued an Announcement of Opportunity for scientific groups to submit proposals for experiments to be performed with a focusing x-ray telescope, and more meetings were held to define the proposed mission. This continued for another year, and another, until in early 1970 the scientists and engineers decided they had a workable design. By this time NASA was ready to go forward with HEAO. They issued an Announcement of Opportunity for proposals for experiments to be flown on one of a series of HEAO missions. In anticipation of this announcement, scientists at AS&E under the direction of Giacconi and Herbert Gursky, at Columbia University under Robert Novick, at MIT under George Clark, and at Goddard under Elihu Boldt, formed a consortium. In

May 1970 they submitted a single proposal as a team, with Giacconi designated as principal investigator.

The integrated facility proposed by the consortium was called the Large Orbiting X-Ray Telescope. It had two x-ray telescopes, one with a large area, equivalent to a 30-inch optical telescope, and low resolution (10 arc seconds) that could respond to high-energy x-rays, and one that responded to low-energy x-rays, had high resolution (2 arc seconds), and an area about five times smaller. In addition, it had a number of x-ray detectors mounted on a wheel so that they could be rotated into and out of the focus of the telescopes.

NASA received proposals from virtually every group studying cosmic rays, gamma rays, or x-rays in the United States, and many from foreign scientists. From these proposals, four missions were formulated. The first two would consist of a mixture of x-ray, gamma-ray, and cosmic-ray experiments. The third mission would be the Large Orbiting X-Ray Telescope, and the fourth would carry a group of cosmic-ray experiments. However, funding was approved for only the first two missions. NASA still had its doubts about the unconventional technology of the x-ray telescope; funds were committed for further study, but not for flight. At least not yet.

When NASA restructured the entire HEAO program in 1973, a scaled-down version of the program survived. The four planned missions were replaced by three smaller ones, each weighing approximately one-third as much as the original experiments. The total cost of the revised program was set at half the original cost.

By 1973, two significant advances had occurred in x-ray astronomy that worked to the advantage of the x-ray telescope consortium in particular and the HEAO project in general. The X-ray Explorer satellite, *Uhuru*, had been a tremendous success. Its data had revealed the nature of the x-ray stars, provided strong evidence for the existence of black holes, and extended the scope of x-ray astronomy beyond the galaxy by detecting active galaxies, a quasar, and clusters of galaxies. Clearly x-ray astronomy was here to stay as a tool for understanding the universe. During the same period, two independent x-ray telescope experiments aboard *Skylab*, directed by Giuseppe Vaiana of AS&E and James Underwood of Goddard, had obtained stunning x-ray images of the sun. Al Schardt, who was then chief of the Astrophysics Division at NASA, remembers: "the *Uhuru* results and the *Skylab* pictures made a tremendous impression; it was clear that x-ray astronomy was important, and it was clear that an x-ray telescope could work. Before that we had never been sure." When the x-ray telescope consortium submitted a proposal for a scaled-down x-ray telescope observatory, NASA approved it, and the x-ray telescope was moved up to be the second of three HEAO missions, following an x-ray all-sky survey. In 1973 Giacconi accepted an offer to become an associate director of the Harvard-Smithsonian Center for Astrophysics, and he moved his group there to form the High-Energy Astrophysics Division. Several factors influenced his decision. The AS&E management was moving more and more toward commercial enterprise; it would have been increasingly difficult to insulate the Space Research Division from this development and to maintain the atmosphere that had made AS&E a unique research institution. Giacconi felt that the Center for Astrophysics would provide a better intellectual milieu for his staff as they prepared for their next major effort, the x-ray telescope observatory. This project was expected to bring x-ray astronomy into the mainstream, and contact with astronomers having a wide range of specialties would be most beneficial. Because it offered the possibility of such interaction, and because of its international reputation as a center of astronomical research, the Center also seemed like an ideal place from which to direct an x-ray observatory that was intended to serve the entire astronomical community. When Giacconi and his group moved, the consortium was expanded to include Harvard-Smithsonian.

The new x-ray telescope observatory, or *HEAO-2* in NASA terminology, had only one telescope of 0.6 meters diameter, a smaller version of the high-resolution, low-energy telescope. Still, Giacconi and his colleagues maintained their hope that the observatory would be the x-ray equivalent of the best radio and optical telescopes. To achieve this would require a complex of instruments able to detect sources a thousand times fainter than those picked up by previous x-ray detectors. No single detector could make detailed x-ray images, take detailed spectral or polarimetric measurements, and also have a large field of view and sufficient sensitivity to detect distant weak sources. To solve this dilemma, the observatory was constructed so that four different and complementary detectors could be rotated into the focus of the telescope. These detectors were the high-resolution imager, the imaging proportional counter, the solid state spectrometer, and the focal-plane crystal spectrometer.

The high-resolution imager, developed at Harvard-Smithsonian

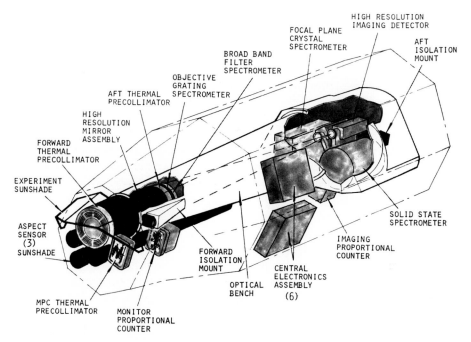

The experiment configuration in *HEAO-2.* (H. Tananbaum, Harvard-Smithsonian Center for Astrophysics)

by Steve Murray, Patrick Henry, Edwin Kellogg, Harvey Tananbaum, and Leon Van Speybroeck, could make detailed x-ray images with 2-arc-second resolution, comparable to the performance of a large optical telescope, but it had a small field of view and could not measure the spectrum of the incoming x-rays. The imaging proportional counter had a wider field of view, could detect weaker sources, and had modest energy and spatial resolution, so it could make broad-brush x-ray pictures. It was designed and developed by Paul Gorenstein and F. R. Harnden of Harvard-Smithsonian. The solid state spectrometer, developed by Elihu Boldt, Stephen Holt, and Robert Becker of Goddard, could make fine spectral measurements over a broad range of x-ray photon energies. The focal-plane crystal spectrometer, developed by George Clark, Claude Canizares, and Tom Markert of MIT, could study very fine spectral details over a narrow range of energies, for strong sources. The x-rays would be focused onto one or the other of the detectors by an x-ray mirror assembly of four nested pairs of highly polished cylinders, each about two feet in diameter. A polarimeter, consisting of three fixed crystal panels, was located in the forward section of the spacecraft and coaligned with the telescope. This instrument, which was designed by Novick and his colleagues at Columbia University, would measure the polarization of the x-radiation, a parameter that gives information concerning the geometry and other conditions in the region where the x-rays are produced. In addition to the primary instruments, the observatory would have two auxiliary instruments. The objective-grating spectrometer could be used with the high-resolution imager on strong sources to obtain spectral information from different parts of an extended source; the monitor proportional counter viewed parallel to the axis of the telescope and would continuously monitor, over a broad range of energies and field of view, the same sources viewed by the telescope.

It was a good design. The question now was, could it be built within the budget and on time? Certainly it would not be a trivial task. The existing technology would have to be pushed to the limit, and in some cases, such as for the high-resolution imager, new technology would have to be invented. Operating within the stringent financial and time limitations required struggles at all levels. If one part of the package slipped in time, it all slipped. If one subcontractor came in over budget, the money might have to be taken away from the funds earmarked for another part of the observatory; if this happened too many times, an instrument might have to be deleted from the observatory altogether, and the scientific integrity of the experiment as a whole might be threatened. A strong, sometimes bruising advocacy system developed. NASA pushed hard to keep the project on schedule and on budget. The scientists pushed hard for the necessary time and money to maintain the integrity of the experiment in the face of increasingly strict budgetary and scheduling difficulties brought about by subcontractors' delays and overruns and by inflation. In the heat of these conflicts, some scientists expressed the viewpoint that NASA managers would fly a bag of sand, if only it was launched on budget and on schedule; NASA officials, meanwhile, felt at times that the scientists were living in a dreamworld where unlimited time and money were available to build the perfect instrument. The polarimeter experiment was an early casualty of the cost-cutting.

One cost-cutting procedure adopted by NASA on all the HEAO

missions was to build only one copy of the complex observatories. The idea was to solve the problems as they developed, not to build or test prototype models of the instruments. This concept, initiated by Richard Halpern, HEAO program manager at NASA head-quarters, and Fred Speer, the HEAO program manager at Marshall Space Flight Center, was called "protoflight," for prototype plus flight. Protoflight was a great cost saver, but it was terrifying for the scientists to contemplate having to build an entirely new instrument such as an x-ray mirror so that it would work correctly the first time it was used.

Protoflight also went against one of the cardinal principles of sound design philosophy, that of "soft failure." In a soft failure, the malfunction of a single instrument may reduce the effectiveness of the mission but does not endanger a substantial portion of the mission's objectives. In a hard failure, such as when the rocket blows up on the launch pad or the spacecraft fails to achieve orbit or an electrical short circuit causes the entire experiment to go dead, all is lost.

Obviously, hard failure is to be avoided, or at least made highly improbable. This is done by incorporating redundancy into the design whenever possible, and by having a prototype ready as a backup in case of a launch failure. It was not possible to have a backup in the protoflight procedure, so all the scientists could do was to hope that the rockets would not fail and take with them years of work and millions of dollars. The principle of soft failure could be exercised to a degree, however, in the design of the instruments. For example, redundant instruments were built in whenever possible, and critical spacecraft and support systems were redundant.

Another basic principle, agreed upon at the outset, was that the scientists would work as a team. This sounds obvious: after all, x-ray astronomy, unlike optical astronomy, had never been done by individuals; it had to be done by teams. Yet, in the era of rocket astronomy, these teams had developed a strong sense of individuality and a keen sense of competition with the other groups. They competed for funds and for the scientific rewards and were at times sharply critical of one another. *Uhuru* had been a one-group satellite involving only the AS&E team. *SAS-3* had been a one-group satellite involving only the MIT team. Other satellites such as the *Copernicus* satellite, the Orbiting Solar Observatories, *Skylab*, and *HEAO-1*, had involved several groups, but they had remained independent, merely sharing space on the satellite like so many tenants in an apartment building. Each group built its own instruments and each had exclusive rights to the data from those instruments.

For the second High-Energy Astronomical Observatory *(HEAO-2)*, this approach clearly would not work. Given the complexity of the observatory, which would be breaking new ground in technology, and the high degree of cost control required, a multitude of problems would inevitably arise during construction. Trade-offs and compromises would have to be made weekly, and everyone would have to come to an agreement as to the best course of action. In other words, the groups could not act as separate tenants sharing space on the satellite. They would have to act as a family.

This was recognized early in the planning phase, when the x-ray telescope consortium of scientists from AS&E, Columbia, Goddard, and MIT was formed in 1970. The members of the consortium agreed that the best way to proceed was to develop a complement of focal-plane instruments capable of achieving the common objective, namely an observatory on a par with the best optical and radio telescopes. This meant that, although the individual groups in the consortium would be responsible for the individual instruments, they would have to design and build those instruments in a manner consistent with the common good. In return, all scientists in the consortium would share the data from all the instruments.

For the unified scientific approach to work equitably and efficiently, considerable thought had to be given to mission operations, that is, to the detailed scheduling of the observations and to the control and monitoring of the satellite. The observations had to be scheduled so that the satellite would move smoothly and efficiently from one target to another; to point the satellite first this way and then that way without regard for the time and energy required to move the spacecraft would enormously reduce the amount of time the observatory was actually collecting data, and hence would reduce the amount of science that could be done with the spacecraft during its limited lifetime. Ethan Schreier, who had worked on similar problems as a member of the *Uhuru* team, was put in charge of mission operations about three and a half years before the anticipated fall 1978 launch. One of his responsibilities was to develop a schedule for the observations to be made with the observatory, taking into account the requests of the scientists in the consortium and the restrictions on the motion of the satellite.

Another aspect of mission operations that required detailed attention was, simply stated, how to talk to the spacecraft and how to listen to it. In other words, the scientists needed efficient ways to command the satellite and to handle the stream of data flowing in from it. These functions are called command and telemetry. As the instruments were developed, command and telemetry data bases had to be created that were compatible with the requirements of the spacecraft control center at Goddard. The process involved continuous trade-offs in the design and limitations of the instruments. For example, each time an instrument was modified because of budgetary constraints or to solve a technical problem, the planning of mission operations had to be modified.

The interaction among management decisions, the modification of spacecraft hardware, and mission planning is illustrated by the controversy over gas thrusters versus magnetic torquing. The plan was that the orientation of the HEAO spacecrafts would be controlled by a set of rotating wheels, which would provide the desired orientation by small changes in their rate of rotation. Over time, external forces such as gravity and the earth's magnetic field would cause the satellite's angular momentum to become so large that the wheels could not change its orientation. This excess momentum was to be counteracted, or dumped, by gas thrusters: a jet of gas sent out to the right, for example, would cause the spacecraft to move to the left. This is a simple, relatively inexpensive, and reliable method for controlling a spacecraft. The only difficulty is that when the gas runs out the game is over. The spacecraft can no longer be controlled, and the perturbing forces will soon send it tumbling out of control. Many space experiments have a limited useful life; the gas supply in the detectors may be exhausted, for example, or fatigue may cause electronic parts to wear out, the gyroscopes used to help the spacecraft navigate may fail, and so on. Furthermore, when a field is evolving rapidly, a set of instruments may be made obsolete by new technology, or by discoveries of new phenomena that may require new observational techniques. In these cases it matters little that the gas-thruster control system builds obsolescence into the mission.

In the case of the *HEAO-2* x-ray telescope observatory, however, it would be ten years or more before another equivalent observatory would be launched. Giacconi, on behalf of the consortium, set to work to have the gas-thruster system replaced by a magnetic torquing system. This system, which had been used successfully on *Uhuru* and other satellites, uses electrical power to magnetize a large "torquing coil." The interaction of this magnetic coil with the magnetic field of the earth can be used to maneuver the spacecraft. Since electrical power is generated by large solar panels on the satellite, it is in inexhaustible supply, and the lifetime of the mission can be greatly extended. The problem with the magnetic torquing system is that it takes time to work. In an emergency, this could be catastrophic. For example, an erroneous command might point the solar panels away from the sun; if the panels could not be pointed back toward the sun within five hours, the batteries on the spacecraft would be discharged and the spacecraft might fail. The magnetic torquing system by itself might not be able to prevent such a hard failure; it would have to be used with gas thrusters as backups.

Giacconi argued that a magnetic torquing system could be added for less than half a percent of the total cost of the mission, an additional expense that was well worth it, in view of the unique nature of the x-ray telescope observatory, which could be expected to provide useful scientific data for several years. The request was not granted, however. The scientists set to work to devise another tactic to prolong the lifetime of the observatory.

Schreier and Tananbaum, who was the scientific program manager for *HEAO-2*, together with Marshall Space Flight Center engineers Tom Recio and Tom Guffin, came up with a strategy for minimizing the consumption of the gas used to power the thrusters. This strategy became known as "momentum management." The satellite will tend to build up spin, or angular momentum, because of perturbing forces. The most important of these is called the gravity gradient force. Because in general one end of the satellite will be closer to the earth than the other end, the gravitational pull will be slightly stronger on one end than the other: about one-ten-thousandth of a percent stronger. This exceedingly small imbalance can lead to large problems, since there is no friction to oppose the gravity gradient force, and since the satellite's scientific work requires a pointing accuracy of a few thousandths of a percent. The gas thrusters provide the opposing force to stabilize the spacecraft and prevent the buildup of angular momentum.

The momentum management team's strategy was to choose targets in a sequence that would tend to offset the buildup of angular momentum. That is, if the spacecraft developed a slow spin in the clockwise direction, they would select a target in that region of the sky. The next target chosen would be one that required turning the satellite in a counterclockwise direction; they would change the

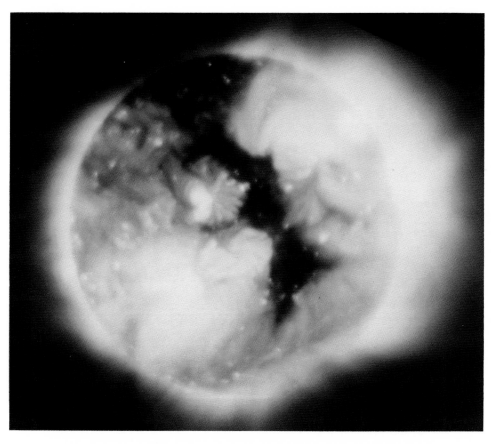

An x-ray photograph of the sun, made by the AS&E x-ray telescope aboard Skylab. (L. Golub, Harvard-Smithsonian Center for Astrophysics)

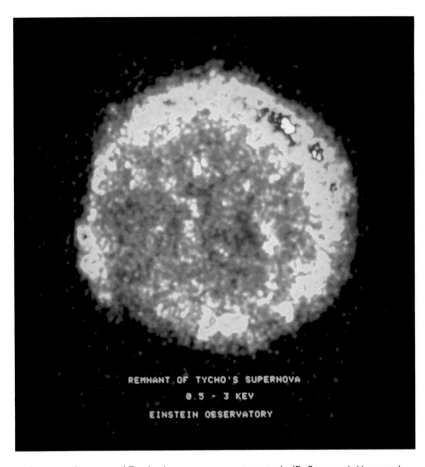

REMNANT OF TYCHO'S SUPERNOVA
0.5 - 3 KEV
EINSTEIN OBSERVATORY

An x-ray image of Tycho's supernova remnant. (F. Seward, Harvard-Smithsonian Center for Astrophysics)

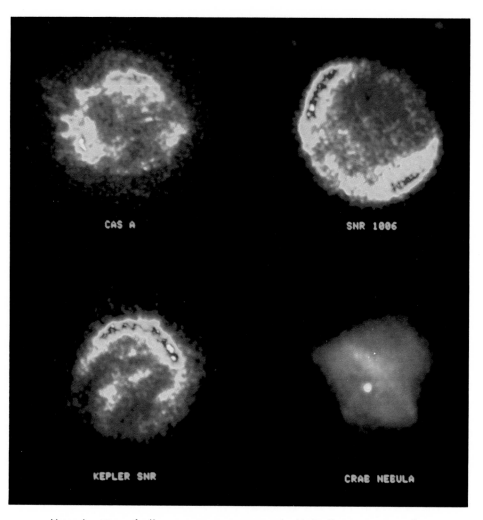

CAS A

SNR 1006

KEPLER SNR

CRAB NEBULA

X-ray images of other supernova remnants. Note the presence of a strong central source, the pulsar, in the Crab Nebula. (F. Seward, Harvard-Smithsonian Center for Astrophysics)

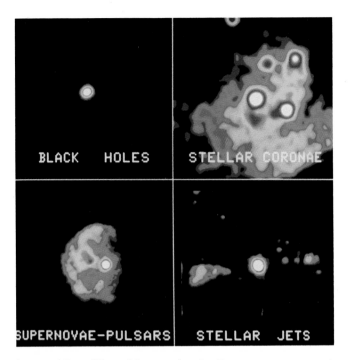

Some of the different types of galactic x-ray sources observed with the *Einstein* observatory: top left, Cygnus X-1, the possible black hole; top right, stellar coronae in the Carina Nebula; bottom left, a pulsar and the surrounding supernova remnant; bottom right, x-ray jets on either side of the x-ray star SS433. (W. Forman and C. Jones, Harvard-Smithsonian Center for Astrophysics)

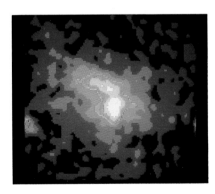

An x-ray image of the galaxy M86 in the Virgo cluster shows hot gas being stripped from the galaxy as it moves through the cluster. (W. Forman and C. Jones, Harvard-Smithsonian Center for Astrophysics)

speed of the reaction wheel, and the satellite would start to drift counterclockwise, and so on. Thus the satellite would bounce back and forth between targets, balancing the perturbing impulses so that the gas thrusters would be needed only occasionally. Another way to save gas was to select targets that had a favorable orientation with respect to the orbit of the satellite. Momentum management would almost double the useful lifetime of the satellite.

As this example illustrates, once a decision had been made for better or worse, all parties concerned, at NASA, TRW, and the science consortium, pitched in to make the best of the situation, and they often came up with brilliant solutions. Without this sort of cooperation the mission could not possibly have succeeded. This example also shows the sometimes double-edged nature of cost-effectiveness decisions. Developing the computer programming and software required for momentum management cost much more than a magnetic torquing system. (NASA did not object because the money came not from the hardware budget but from postlaunch operations funds; these funds could be used at the discretion of the consortium for a variety of purposes relating to mission operations, such as momentum management or data analysis.) This does not take into account the incalculable loss of scientific data due to the premature demise of the observatory. It will be another decade before the United States puts a second x-ray telescope in orbit.

While Schreier, Tananbaum, Giacconi, and others were struggling with problems of mission operations, other scientists were in the basement of the Harvard-Smithsonian Center for Astrophysics, working on the high-resolution imager. All the detectors were pushing the state of the art to the limit, and each presented its own special problems: the high-resolution imager, unlike the other detectors, had no previously existing prototype that had been tested in other applications.

Over the course of five years, Tananbaum, Murray, Henry, Kellogg, and Van Speybroeck worked on the high-resolution imager. Early in the program, the major effort went into trying to develop a new type of detector called the NEAD, for negative electron affinity device. Finally, the group realized that the NEAD could not be made to work within the cost and time limitations, and they gave up on it. They then turned to another type of detector, which uses an array of microchannel plates.

A microchannel plate array is a collection of millions of glass tubes that are coated with a material that converts x-rays into electrons. When an incoming x-ray strikes that material, it liberates an electron. The electron continues down one of the tubes, striking the wall and producing more electrons, until an avalanche of millions of electrons emerges from the bottom of the array. This spreads across a small vacuum gap, where it encounters a grid of wires and produces an electric signal. The trick is to design the grid so that the electric signals can be decoded and you can figure out exactly where the x-ray hit the microchannel plate. The group tried to make a grid consisting of alternating strips of conductors and insulators about a hundredth of a millimeter wide. Early models tended to short-circuit easily and presented other difficulties as well. While searching the literature for ideas, Kellogg came across a description by scientists at the University of Leicester of a type of detector that sounded promising. Murray and Giacconi suggested that a modification of their technique might work. This device made use of the fact that in x-ray astronomy the photons from a strong source arrive at a rate of about a thousand per second, whereas in optical astronomy the typical rate runs into the millions per second. When the counting rates are low, clouds of electrons produced by individual x-rays can be measured. An electronic processor can be used to measure the center of the charge cloud, even if the wires are a few tenths of a millimeter apart. This centroid gives an accurate measurement of the location of the incoming x-ray. Measuring the location of many x-rays makes it possible to reconstruct a detailed image of the x-ray source with wire spacings well within the technological capabilities. Such a detector would not be suitable for an optical telescope, since optical count rates are generally so high that the clouds would overlap, and reconstruction of the image would be impossible.

The Smithsonian group set to work using the basic idea of this prototype to build a high-resolution imager. Murray and Van Speybroeck spurned the suggestion that they contract with some high-tech company to build the device. Instead, Murray decided to build it in the basement with a graduate student so they would know everything there was to know about it and how it worked. The problems were basic: how to make the wires in the grid the same distance apart, how to hold them together, what are the right dimensions, and so on. Their approach was simple. In Murray's words, "We made no engineering studies, no estimates as to what was the optimum size. We just said, 'Let's build something and then try to learn from that how to make it better.' It took a few years, going down many blind alleys, but finally, sometime in 1975, we succeeded. Using the simplest possible equipment and a little bit of glue, we made a workable grid about a half-inch across." They subsequently tried more elaborate designs, but these worked no better. The simplest design was the best, and the original basement design was used on *HEAO-2*.

Problems in testing the high-resolution imager took months to unravel. The team could not understand what was causing the difficulties. Was it the detector, or was it the testing apparatus? Two scientists from the University of Leicester group, Kenneth Pounds and Kenton Evans, brought a prototype to Harvard-Smithsonian. At first their unit did not produce sharp x-ray images. After they ironed out problems in a computer program and a vacuum pump that was feeding electrical noise into the system, the unit worked. This gave the Harvard-Smithsonian group confidence that the problem was in the testing equipment. After a few weeks their detector worked as well as or better than the Leicester one.

The question remained: how would the imager work in space? The primary unknown was how the charged particles in the upper atmosphere would affect its performance. Prototypes of the detector were built and flown on rockets. On the first flight a high-voltage relay failed. On the second flight the pointing system of the rocket payload failed. On the third flight the doors to the rocket did not open. On the fourth flight a switch did not work. Finally, four months before launch, the detector was successfully tested in a rocket flight.

A working high-resolution imager was clearly essential for the optimum performance of the observatory; without it no detailed x-ray images could be obtained. Likewise, the imaging proportional counter, the solid state spectrometer, and the focal-plane spectrometer had key roles to play. At the same time, if any one of these instruments failed the mission would not be a disaster; much useful science could still be done with the other instruments. For example, if the high-resolution imager failed, the imaging proportional counter could still make images that, although not as good as the high-resolution images, would nevertheless be better than any obtained before, and detailed, highly sensitive observations of many objects could still be made. If the mirror assembly failed to perform, however, the mission would be a failure. It was an essential component of the observatory, and no one could be sure that a high-resolution mirror assembly of the type proposed could be made to work. The largest previous high-resolution mirror was the one flown by the AS&E team on *Skylab*, which had about one-tenth the collecting area of the one proposed for *HEAO-2*. Not surprisingly, the mirrors received more attention than any other single item.

Leon Van Speybroeck had primary responsibility for developing the mirror assembly. He had been working since 1967 to develop x-ray mirrors for the solar x-ray astronomy program, including those for the solar x-ray telescope flown on *Skylab*. He and his colleagues had to determine the optimum design for the mirror assembly, in other words, the design that would achieve the highest resolution for the lowest cost. Along with Richard Chase of AS&E, Van Speybroeck developed an elaborate computer program that would give the theoretical response of different mirror designs. In the laboratory he conducted experiments to measure the scattering of x-rays from polished flat mirrors made of different materials. These experiments gave the group confidence that achieving resolution of one arc second, a value comparable to that of the best optical telescopes, would not be a fundamental problem. It could be done with fused silica, the material from which most glass is made, with a thin coating of evaporated chromium and nickel to improve the efficiency of the x-ray reflections.

The final design involved four concentric sets of nearly cylindrical surfaces, each about two feet in diameter. The interior surfaces of the mirrors were to be the shape of a parabola followed by a hyperbola. The curves would be so subtle that the mirrors would resemble cylinders with a slightly conical internal surface. X-rays entering the telescope would encounter the mirror at a grazing angle, and would reflect first off the parabolic portion of the mirror and then off the hyperbolic portion. These two reflections would focus the x-rays to a point. Four such mirrors would be nested inside one another, like successively smaller measuring cups.

Going from the design to the reality of a working mirror assembly was not easy. Since it was a unique design, many things were being done for the first time, and problems had to be solved as they came up, always under tremendous pressure from already nervous NASA personnel who were themselves under pressure to keep the project on budget. If they fell too far behind schedule or went too far over budget, the mission might be canceled. Van Speybroeck recalled: "the scientific community and NASA thought that this kind of mission was verging on the unfeasible. *HEAO-1* was based on the supposition that *HEAO-2* might fail."

One of the first problems was to find a manufacturer who could make large glass cylinders of the desired shape and quality. The seams in the mirrors could not have bubbles. After much searching, a German company, Heraeus Schott Quarzschmelze, came up with a workable method. They would make the cylinders in two steps. First they would make a series of glass "barrel staves" of the right shape. Then they would fuse them together in a centrifuge. This involved process took almost a year, but it worked.

The next step was to use these cylinders to make mirrors. This was done at Perkin-Elmer Corporation in Connecticut by a team headed by Peter Young. The cylinders were first diamond-ground to the approximate shape of paraboloids and hyperboloids. Immediately another problem arose. The nested configuration required thin cylinders — so thin that they would sag under their own weight in the earth's gravity. This made it virtually impossible to determine whether the mirrors had been ground to the correct shape. Although the laboratories where the mirrors were fabricated were cleaner than a hospital operating room, still there would be tiny specks of dust on the measuring table. These specks would distort

The x-ray mirror assembly. (Perkin-Elmer Corporation)

the mirror, causing an inaccurate measure of its roundness. The use of rigid support structures also caused changes in the shape of the mirror. A way had to be found to simulate the free-floating state of space. It occurred to Van Speybroeck that a person floating in the water is in an almost weightless state; the mirrors would not float in water, but they would float in mercury. He contacted NASA for permission to try it. NASA was skeptical. Van Speybroeck decided to try it on a small prototype mirror. Ed Fireman, an astrophysicist at Harvard-Smithsonian, had some mercury left over from an experiment; he made it available to Van Speybroeck and his colleagues, who showed that they could accurately measure the roundness of the mirror by floating it in mercury. NASA then approved the procedure for the actual flight mirrors.

New difficulties appeared around every corner. The next step was to polish the mirrors to telescopic quality, that is, to a smoothness of about one-ten-millionth of an inch. This is done by polishing awhile, testing for roughness, making corrections, and polishing

The first model of the high-resolution imager. (American Science and Engineering)

some more. The problem was that there was no easy way to test the mirror with x-rays. The theory of how x-rays scatter from rough surfaces was poorly developed, and the techniques used at optical wavelengths would not work for nested x-ray mirrors. The group knew that a small imperfection in the surface finish of the mirror could cause a large loss in performance, but it was impossible to make accurate estimates of how small the imperfections were and how large the loss in performance would be. The approach was to polish the mirrors as well as was possible with the existing techniques; this led to complaints from NASA that the scientists and engineers were being overprecise and were, as a result, running up the budget and falling behind schedule. Ted Kirchner of AS&E worked with Young to rearrange the schedule to provide extra hours for polishing. Eventually, though, the NASA program office, nervous because the project had fallen a few days behind schedule and worried that the untried procedures for coating and alignment might cause further slippage, directed Perkin-Elmer to stop polishing the mirror one day early and proceed to the next step of coating the mirrors and aligning the assembly. This decision eventually caused a substantial loss of efficiency at high energies, where it was needed most because the mirrors naturally become less efficient at high energies.

The coating went relatively uneventfully. Sample mirrors that were coated at the same time showed no additional imperfections due to the coating. Assembling and aligning the mirrors presented yet another set of problems. The mirrors had to be supported by a substructure made of other materials. This structure had to support the mirror without much distortion during ground testing, protect it during the intense vibrations of launch, be noncontaminating, be stable, and not conduct heat into the mirror. These requirements were met with a combination of graphite-epoxy cylinders and Invar (an iron-nickel alloy) flanges designed by AS&E engineers. While the mirrors were bonded to this structure and aligned, they had to be suspended so that they would assume the same configuration they would have in the weightless conditions of orbit. Under Van Speybroeck's nervous, watchful eye, Gerry Austin and William Antrim of AS&E, along with the Perkin-Elmer engineers, devised an elaborate system of thirty-two counterweights acting on each mirror so that they hung in a weightless state while the entire assembly was bonded together and aligned.

By early 1977 the mirror assembly was complete. During the

Assembly and alignment of the x-ray telescope at Perkin-Elmer.
(Perkin-Elmer Corporation)

two-year process of fabrication at Perkin-Elmer, the Harvard-
Smithsonian, AS&E, and Marshall groups were present almost continually; this constant consumer presence must have been a burden
to the Perkin-Elmer staff. To their credit, they bore up under it
gracefully. This type of cooperation between contractors and the
consortium was a key element in the success of the project, because
as questions arose — and there were many — they could be answered on the spot. After fabrication, the mirror was to be shipped
to AS&E in Cambridge, where the entire scientific package would
be assembled and tested. From there it would go to the Marshall
Space Flight Center in Huntsville, Alabama, for vacuum testing.
Then it would be shipped to TRW, near Los Angeles, where it would
be fitted to the spacecraft to complete the observatory. After this
procedure, called integration, the observatory would be shipped to
Cape Canaveral for the launch. A nearly fatal error occurred even
before the mirror assembly got out of the Perkin-Elmer parking lot.
The crate containing the mirror was to be loaded by a crane onto a
truck. Luckily, the ever cautious scientists and engineers insisted
that this process be tested first in a dry run, with a crate containing
ballast. The crane dropped the crate! After some checking and re-
checking of the crane, it was used on the mirrors, and it worked. The
mirror assembly was safely on its way.

At Marshall Space Flight Center the entire observatory would receive its first x-ray test. All the parts had been tested and shown to work individually. They had been put together to work as a unit at AS&E. But would the package work as an x-ray observatory in the vacuum of space? Apart from this fundamental question, the observatory had to be calibrated. The word *calibrate* comes from a Greek word meaning "wooden foot," that is, the carved block that shoemakers used to make shoes for a customer. In more recent times the word came to refer to the determination of the calibre, that is, the inside diameter, of a tube, such as a gun barrel. If you had a set of rods that would fit snugly into gun barrels of known calibre, then you could calibrate other gun barrels by testing to see which rod, corresponding to which calibre, would fit into each barrel. In the same manner, the x-ray observatory would be tested under known conditions. For example, an x-ray source of a known intensity and spectrum would be placed at a known angle in front of the telescope, and the response of the telescope would be measured. This would be done for a large variety of angles, source strengths, and so on, to test the response of the observatory over a wide range of conditions.

In preparation for the calibration of the *HEAO-2* x-ray telescope, a team of scientists and engineers from the consortium, Marshall Space Flight Center, and the industrial subcontractors, AS&E and TRW, designed and built a special x-ray calibration facility at Marshall. This facility includes a 1000-foot-long pipe, which makes it one of the most striking of many unusual facilities at Marshall. The long pipe was necessary to allow effective focusing of a small x-ray source at a large distance from the telescope. The inside of the pipe was kept at a partial vacuum, so the x-rays would not be absorbed by atoms in the air between the x-ray source and the telescope. At the end of the tube opposite the x-ray source was a huge vacuum chamber (20 feet in diameter, 40 feet long), where the telescope would be housed under conditions similar to those it would encounter in orbit.

The original schedule called for six months of testing. This would give adequate time to test the observatory at several x-ray energies, to move the source so that the telescope viewed it at many different angles, to change the temperature in the chamber to at least three different values, and to troubleshoot if any problems should arise. The first problem that arose was the schedule. By the time the observatory arrived at Marshall, the schedule had already slipped

The HEAO-B x-ray test and calibration facility at Marshall Space Flight Center. (NASA)

The 1000-foot pipe for focusing x-rays. (NASA)

six months. To make up some of this time, NASA shortened the testing time to one month. Giacconi protested loudly. More than a thousand measurements were needed to make sure the observatory was ready for flight. Even if the average time per test could be cut to half an hour, including the time required to change the source or move it around, change the temperature, and change the instrument configuration, a minimum of five-hundred hours would be needed. This would take thirty-one days, working double shifts of sixteen hours a day, and making no allowances at all for the inevitable problems that would come up. The NASA program office held firm. Thirty days. Giacconi agreed on the condition that the scientists of the consortium be placed in charge and that all the teams involved, from TRW, Marshall, and so on, work under the direction of the consortium.

Realizing that every minute of the calibration testing was precious, Giacconi directed Van Speybroeck to develop a computer program to make the best possible use of time during the testing. They sought the best possible sequence of testing, along with accurate estimates of the total time required for each test. A schedule of

1397 different tests was generated, with an average time for each test of 18 minutes and 30 seconds. This could be done by working 16 hours a day for 27 days, or 24 hours a day for 18 days. They chose the 24-hour-day schedule, with overlapping 13-hour shifts. This would allow 12 valuable days for solving problems and retesting. Fred Speer, the HEAO program manager at Marshall, agreed to provide the staff support to keep the calibration facility running 24 hours a day. It was a tense, hectic time, a time filled with surprisingly little acrimony considering the tremendous pressure that was bearing down on everyone. One dramatic moment came when the high-resolution imager was rotated into the focus of the telescope for the first time. Would the mirror and the imager work as designed? Van Speybroeck, who had experienced a queasy stomach before in anticipation of this moment, began to feel ill again and chose to remain in Cambridge. The x-ray source was turned on. The readout from the high-resolution imager was checked. It showed nothing, except perhaps a weak diffuse signal. Certainly not the bright pointlike image there should have been. Was the telescope a failure? Giacconi and Tananbaum exchanged nervous glances as Giacconi ordered the intensity of the source increased. Still nothing. They increased it further. Still nothing. They shut off the x-ray source. What was wrong? They began to go over all the possibilities, most of them too dismal to contemplate. Finally they remembered the ribbon. In order to conduct some special tests, they had placed a thin ribbon in front of the high-resolution imager. Did the x-ray image by chance fall on this ribbon? If so, just tilting the telescope would uncover the image. They repeated the test, with the telescope tilted. This time, like a rising sun, a beautiful pointlike image appeared. The telescope was working exactly as designed. Over the next four weeks the scientists would experience many nerve-wracking moments, but perhaps none so heart-stopping as this one.

Thanks to the calibration team's heroic efforts, the calibration of the observatory was successful. However, as with other cost-cutting crash efforts, the cost-effectiveness of shortening the testing time was questionable. The imaging proportional counter developed a very slow leak that complicated its calibration. This could have been cleared up in a few weeks of testing; instead, the imaging proportional counter has had to be calibrated in orbit, a task that has taken several man-years of work.

The process of calibrating the observatory generated a flood of data. Knowing in advance that it would be essential to analyze these data immediately, Arnold Epstein, William Forman, Josh Grindlay, Christine Jones, Jeffrey Morris, Schreier, Van Speybroeck, Murray, and Eric Feigelson developed an efficient data-handling system. The computer software developed at this time formed the basis for the data-handling system that would be used to reduce and analyze the data once the observatory was in orbit. When a powerful new astronomical observatory is designed and constructed, it often happens that considerable attention is focused, rightfully so, on perfecting the hardware, but the means of handling the data are not given enough attention. The result is a glut of data, and months pass before meaningful scientific results are obtained. This has happened with both ground-based and space observatories. In contrast, planning for data analysis for *HEAO-2* began more than three years before launch. This system was tested along with the instruments during calibration, which produced a volume of data equivalent to more than a month of operation in orbit. The *HEAO-2* data-handling system, one of the best ever developed for a scientific experiment, would allow the scientists to begin analyzing the data and reporting scientific results almost immediately after launch. This was an important consideration for the scientists, who had been waiting years for the x-ray telescope observatory and were anxious to get the results, and for students, such as Feigelson, whose doctoral thesis depended on data from the observatory.

In September 1977, the observatory was shipped to TRW in California. TRW was the contractor for the spacecraft that housed the HEAO experiments. Marshall Novick was the HEAO program manager at TRW, having taken over the job from Richard Whilden, who had guided the program through its design phase. During the several months it took to integrate the experiment into the spacecraft, Patrick Henry of Harvard-Smithsonian stayed at TRW to look after the interests of the scientific consortium. It was TRW's responsibility to build the spacecraft, mate the observatory to the spacecraft, support launch operations, and perform flight operations at mission control center at Goddard Space Flight Center.

The spacecraft equipment module was an eight-sided prism about 1 meter tall and 2½ meters in diameter. It provided the support systems for the spacecraft and the observatory, such as the gas for the thruster system and the electronic components needed to transmit power to the observatory, to relay commands from earth to the observatory, and to relay the data from the observatory to earth. Solar panels would produce about 400 watts of power to run

Checking out HEAO-B in a clean room at TRW. (TRW)

all the equipment on the spacecraft. In the event of a power loss, a low-voltage sensor would automatically command the observatory into a contingency mode in which the solar panels would point toward the sun. This would ensure the survival of the spacecraft until ground control could solve the problem. The overall length of the spacecraft, including the observatory and the equipment module, was about 19 feet, roughly the size of a small truck.

The observatory would be controlled from the mission control center at Goddard. Raw data would be collected and stored on on-board tape recorders until it could be transmitted to tracking sites, whence it would be transmitted back to Goddard. Early in the planning, the scientists decided to transmit raw data to the ground, rather than to process the data on board before transmitting it to the ground. This decision made the analysis of the data on the ground more complex, but it more than paid for itself. It made it possible to relax the pointing accuracy and stability requirements of the spacecraft considerably. The key is to have instruments that detect individual photons and to know accurately where you are looking at each instant of time. To make this possible, the consortium designed an elaborate aspect system that worked by reference to visible stars and an internal calibration system. These systems meant that the telescope did not have to be pointed exactly toward the source during an entire observation, which might last a day, but only long enough to determine exactly where the source was, which would take about one second. Keeping track of where each photon hit the detector and where the detector was looking at each moment made it possible to construct an accurate map of the sky, even if the telescope wandered around slightly during an extended observation. This relaxation of the pointing requirements significantly reduced the cost of the overall project.

By October 1978, the spacecraft at last was ready for launch. It was shipped to Cape Canaveral.

10
Einstein into Orbit

Some of NASA's research satellites have had colorful, evocative names such as *Mercury, Apollo, Pioneer, Viking,* and *Voyager.* Others have had bland, functional names such as *OSO-1, 2, 3, 4,* and so on. The official names for the High-Energy Astronomical Observatories were in this tradition: *HEAO-1, HEAO-2,* and *HEAO-3.* Giacconi felt strongly that this tradition should be broken, that *HEAO-2,* like the *Uhuru* x-ray satellite, would be pushing past the existing frontiers of knowledge, and that like *Uhuru* it should have a unique, memorable name. Harvey Tananbaum suggested *Pequod,* the name of Captain Ahab's ship in *Moby Dick.* Giacconi, whose position as principal investigator of the *HEAO-2* project had inevitably taken on some of the characteristics of a benevolent dictatorship, was well aware of the inevitable comparisons between himself and Captain Ahab that such a name might evoke. Nevertheless, he liked the name. The metaphor was apt. Ahab, in the tradition of Ulysses, had driven his crew in search of a more profound knowledge of the universe. *Pequod* also had a nice crosscultural ring to it, eliciting images of the American Indian and the New England whalers.

Giacconi contacted Noel Hinners, associate administrator for space science at NASA, to suggest the name *Pequod* for *HEAO-2.* Hinners' response was simple and straightforward: No. It would take two years to get the name changed; besides, NASA didn't like the idea of associating one of its satellites with a white whale. Giacconi then called Frank Press, the President's science adviser, to see if the White House was interested in naming the satellite. Press mentioned that both President Carter and his daughter, Amy, were reading Walter Sullivan's book *Black Holes* at the time, so maybe they would be interested. But nothing came of it. Nevertheless, Giacconi was determined to give the satellite a name, even if unofficially. This was how *Uhuru* had been named, after all. If the name was to stick, though, everyone in the consortium would have to like it and use it in publications. Accordingly, all members of the consortium were asked to submit names, a list was prepared, and a vote was taken. The name chosen was *Einstein:* the flight of the spacecraft would honor the centennial of the birth of the man whose theories of space, time, and matter pushed back the frontiers of our knowledge of the universe. Everyone agreed it was an excellent choice. Although the name has never been officially recognized by NASA, *HEAO-2* is now almost universally referred to as the *Einstein* observatory.

As the launch date neared, the observatory was checked and rechecked almost continually, but everything went very smoothly. A team from Harvard-Smithsonian, including Giacconi, Tananbaum, Henry, and Gerry Austin, arrived at Cape Canaveral about a week before launch for a final series of tests and review meetings. The launch vehicle would be the Atlas-Centaur, a rocket built by General Dynamics Convair Aerospace Division under the direction of NASA's Lewis Research Center in Cleveland, Ohio. The Atlas is powered by three Rocketdyne engines — two with 370,000 pounds and one with 60,000 pounds of thrust. The Centaur upper stage is powered by two Pratt and Whitney engines with a total thrust of 30,000 pounds. The total height of the HEAO Atlas-Centaur space vehicle ready for launch was 39.9 meters.

During the final review meetings, some questions came up about the readiness of the launch vehicle. According to Harvey Tananbaum, "the rocket engineers made some casual statement that the flight computer controlling the vehicle had had some problems in the past few days. People weren't quite sure what the problem was. There weren't many spare parts left, either. With each succeeding launch, the Centaur parts got used up. We expressed concern; they reassured us that everything was all right, that they always had problems like this, and not to worry, it was a good reliable launch vehicle. But we did worry, and we kept questioning them closely. That launch vehicle didn't have to be good. It had to be perfect."

By the day before launch, almost all the members of the scientific consortium were there, many of them with their families. Van Speybroeck, whose skittish stomach could not stand it, did not go. Instead he went to Goddard Space Flight Center, where he monitored the prelaunch activities, with a direct telephone link to Giacconi in case any problems should arise. George Clark, who had worked with Giacconi eighteen years earlier on the first estimates of the expected intensity of x-ray sources and who was now the principal scientist for the focal-plane crystal spectrometer on *Einstein*, was there. Bruno Rossi, who had first suggested to Giacconi that x-ray astronomy might be an interesting field and had coauthored with him the first paper describing the design and potential usefulness of an x-ray telescope, was there. In the intervening years he and Giacconi had drifted apart. Clark used the occasion of a prelaunch party given by TRW to bring them together again, as he proposed a toast to "both the father and the grandfather of x-ray astronomy."

The launch was scheduled for a few minutes after midnight on Monday, November 13, 1978. On Sunday afternoon, *Einstein* received its final checkout, and everything looked fine. Singly and in groups, the scientists gathered to watch. Tananbaum, Henry, Austin, and Giacconi took one last drive out to the launch tower to look over the rocket; the spacecraft, perched perilously atop a fifteen-story rocket, looked terribly vulnerable, yet beautiful. They made one last, attentive, affectionate survey and drove back to the grandstands. Giacconi took his place in the control area with NASA officials; he would have the final say on whether to stop the countdown in case questions arose about the readiness of the scientific payload. Next to him was Pat Henry. Henry had followed the experiment from calibration through the integration with the spacecraft at TRW; more than anyone else, he would know if anything was going wrong with the observatory.

The countdown proceeded smoothly down to T minus 20 seconds and counting. At this point Pat Henry knew it was going to go. He gave up his seat in the control room and ran outside to watch. At 12:22 A.M. the tower lit up. "Ignition," came the matter-of-fact announcement over the intercom. With an explosion of sound, vibration, and light, the vehicle lifted off, rising with agonizing slowness at first, then picking up speed. About 4 minutes later, the booster engine separated from the main engine. The main engine burned for a little over 7 minutes, taking the spacecraft more than 200 miles above the surface of the earth. Approximately 23 minutes after launch, and 15 years after the first proposal for an x-ray observatory, the *Einstein* observatory achieved orbit. Only then could Giacconi relax. He joyously hugged his 15-year-old son, who had come to the launch to "see what his old man was up to."

The launch of the *Einstein* observatory. (NASA)

11

First Light

The jubilation over a successful launch was short-lived. It was quickly replaced by the eager, nervous anticipation of "first light," the moment when the instruments would be turned on, the telescope would be pointed toward a particular source, and the spacecraft would transmit to earth its first image of an extraterrestrial object. For the scientists, first light would be the climactic moment. The choice for first light was, appropriately enough, the most famous x-ray star, Cygnus X-1, one of the strongest x-ray sources and a suspected black hole.

Before the observatory could begin operating, a carefully orchestrated procedure had to be followed. First, a few hours after the satellite achieved orbit, the low-voltage power supply would be turned on. If everything checked out all right at low voltages, the high-voltage power supply would be turned on. During this time the star trackers would be calibrated and the spin maneuvers checked. If everything was working, the observatory would swing around to point at Cygnus X-1. During the checkout period, most of the scientists returned to Cambridge to get some much-needed rest and prepare for the flood of data they hoped would soon be on its way. Steve Murray and Ethan Schreier, the two experts on observatory operations, after a sleepless night, caught the early morning flight directly to Goddard, where they would oversee the instrument checkout.

The low-voltage test went off without incident. The high voltage was turned on and everything looked okay, so Murray and Schreier ordered the star trackers to be activated. The star trackers track bright stars and make it possible to determine quickly and precisely where the satellite is pointing. Without them the observatory would be severely handicapped, if not totally incapacitated. Soon the data from the star trackers came reeling out of a computer, printed on a stripchart. As Murray and Schreier pored over these charts, they became increasingly apprehensive. Whatever the star trackers were tracking was advancing across the field of view much faster than the satellite was moving. At the time, the satellite was over the Pacific Ocean somewhere near Hawaii. At this early stage of the mission, the spacecraft had yet to be maneuvered to its proper orientation, and the star trackers were pointed toward the earth. It was nighttime in Hawaii, so it was thought that the star trackers might have picked up the lights of Honolulu. But these lights should have moved across the field of view at the same rate that the spacecraft was moving. The rapid motion of the light through the star trackers made no sense.

Not often, but occasionally, electrical components of instruments under the high-vacuum conditions of space will experience high-voltage arc discharges, or corona, from one component to another, causing the instrument to malfunction. Most space scientists would agree with Steve Murray's assessment: "corona is the most insidious problem in space applications; it is probably the major cause of in-flight failures." Corona caused the failure of the Orbiting Astronomical Observatory on which Philip Fisher of Lockheed had launched what could have been the first operational satellite-borne x-ray detector. One of the main goals of preflight testing is to put the instruments into a high vacuum and test them for corona. Despite the extensive testing of the *Einstein* star trackers, there was, recalled Ethan Schreier, "a possibility of corona in the star trackers. A strong possibility. We ordered them turned off." This brought a howl of protests from mission control. Turning the star trackers off meant delaying the activation schedule, which meant keeping a large staff standing by with nothing to do. Murray and Schreier held fast. They were not sure what was happening, but they preferred to be cautious rather than burn out the star trackers. The star trackers were turned off while the two scientists puzzled over what could be the matter and what they would do if the star trackers really were arcing.

Then, as Schreier tells it, "I don't know when it finally came to us, but we went outside at night, and there was the moon. I remembered seeing the moon at midnight on the night of the launch. It was beautiful. It hit us. We were seeing the reflection of the moon on the

Pacific Ocean." A quick calculation verified that the rapid motion seen in the star tracker could be caused by the rapidly changing angle between the satellite and the moon's reflection. The scientists agreed to turn the star trackers back on. One at a time. They worked. Eventually one of the star trackers did fail, for other reasons, but because of the concept of soft failure the observatory had been equipped with three star trackers, so the failure of one caused no problems.

Four days after launch, all conditions were "go" for putting the high-resolution imager into the focus and pointing the telescope toward Cygnus X-1. Giacconi and Henry joined Murray and Schreier at Goddard for this occasion. They had set up at Goddard a real-time image-processing system so they could observe what the telescope was seeing as it was seeing it. It was one of those joyful, almost intoxicating moments that occur all too seldom in a scientific career. As the telescope scanned toward Cygnus X-1, x-ray

First light from the *Einstein* observatory; an x-ray image of Cygnus X-1. (NASA)

images of stars streaked across the field of view. Then the telescope fixed on Cygnus X-1, and as its pointlike image formed on the screen, the scientists knew they had an x-ray telescope that worked.

The *Einstein* observatory remained operational for about two and a half years, until April 26, 1981, when the supply of gas ran out. During that time it detected thousands of new x-ray sources and accumulated a lode of data that will continue to yield valuable scientific results for years.

The *Einstein* observatory has also brought about a sociological change in x-ray astronomy. It has changed the way things are done, from experiments by individual groups to the shared use of large facilities. It has also made x-ray data available to a broad segment of the astronomical community through the guest observer program. In this program, which has been administered by Fred Seward of Harvard-Smithsonian, a substantial amount of observing time was reserved for guest observers. Because of the large data bank, this program remained in operation long after the observatory ceased to collect data. More than five hundred guest observers from over fifty different institutions around the world have taken part in the program. Thus, *Einstein* brought x-ray astronomy into the mainstream of astronomy not just by virtue of its ability to observe a wide range of astronomical objects but also because it made x-ray data available to a large number of astronomers. In doing this it has helped to propagate the notion that astronomy is best done by looking at the universe over a wide range of wavelengths, from radio to x-ray.

Within only sixteen years, we have progressed from a simple detector comparable to Galileo's telescope to an x-ray observatory comparable to the 200-inch Palomar Mountain telescope. In both cases the increase in sensitivity was roughly a million-fold. When Galileo first turned his newly developed telescope toward the night sky on a winter night in 1609, our perception of the universe changed forever. He saw the moons of Jupiter, craters and mountains on our moon, and thousands of stars invisible to the naked eye. As Galileo's simple but effective instrument was improved upon over the next three centuries and more, until it became the sophisticated giant on Palomar Mountain, more marvels of the unseen universe came into view — dense white dwarf stars, majestic spiral galaxies, and the still-enigmatic quasars.

The history of x-ray astronomy has been similar, though vastly compressed in time. The first x-ray detectors revealed a new face of the universe — a face of extremes, of stars collapsed to form neutron stars and black holes, of the exploded remnants of stars, of galactic nuclei in turmoil, and of clouds of hot gas spreading over millions of light-years. It was a fascinating face, but it could be glimpsed only vaguely because of limitations of the detectors, and x-ray astronomy remained somewhat behind and apart from the mainstream of astronomy. With the flight of *Einstein*, x-ray astronomy joined the mainstream. Almost every type of object known to optical astronomy, from small dwarf stars to quasars, has now been detected and imaged in x-rays, with a quality comparable to that obtained with the major optical telescopes, and many phenomena have been seen that are invisible with optical telescopes. The x-ray universe has come into focus.

On August 28, 1980, an unknown malfunction aboard *Einstein* triggered a failsafe system that automatically shut down all the power circuits, including those of the gyroscopes used to position and point the spacecraft. When the systems were turned on again, only two gyros would operate. At launch the spacecraft had six working gyros, double the number needed. This was thought to be more than enough redundancy, but one failed several months after launch and a second functioned only sporadically. Now two more had failed. With only two gyros, the spacecraft began to tumble out of control. It was brought back into control by using the gas thrusters. This method gave only crude pointing accuracy and used large amounts of the precious gas, but it allowed the group to buy time to see if they could fix the gyros. Rescue teams from Harvard-Smithsonian, TRW, Goddard, and Marshall Space Flight Center set to work. By working night and day to rewrite computer programs so that the satellite could use the sun and star sensors together with the two good gyros to fix its orientation, they were able to keep the satellite functioning until December 5, when one of the dead gyros started to operate again. Although about ten months' worth of gas had been used, the rescue teams' heroic efforts extended *Einstein*'s lifetime by more than four months. On April 26, 1981, the gas supply was exhausted, and the *Einstein* observatory ceased operation. During its lifetime, it made more than five thousand separate observations of objects ranging from comets to clusters of galaxies.

In the chapters that follow, we give a sampling of the results from *Einstein*. We make no attempt at completeness; such an attempt would be futile, since important new results are continually coming to light as astrophysicists work their way through the rich *Einstein* data banks. Rather, we touch on some areas of research that interest us personally, and in which we believe *Einstein* has had an especially significant impact. Even this limited survey describes the contributions of scores of scientists; because of their large number, we regrettably cannot mention all of them by name or cite their work. For these important details, we refer the reader to the books and journals listed in our Bibliographical Notes.

12 Stellar Coronas and Supernovas

Among the first observations scheduled for the *Einstein* observatory were deep surveys. The idea was to look at blank fields, regions empty of known radio, optical, or x-ray sources, for a long time, say an entire day, so as to get a deep exposure, and to see what, if anything, showed up. In almost every case, two types of objects were found. There were quasars, incredibly luminous objects that appear as faint x-ray sources only because of their enormous distances of billions of light-years, and there were normal stars that are among our nearest neighbors in the galaxy. These stars are some of the faintest sources of x-rays detected outside the solar system. Still, many of them are much brighter in x-rays than could be predicted from what was known about the x-rays emitted by the sun; for this reason, far more normal stars were found to be x-ray sources than had been expected. In nearly every extended observation, some nearby stars show up in the field of view.

The full value of this unexpected bonanza has yet to be appreciated, but it is already clear that *Einstein*'s observations of the x-ray emission from normal stars will be an extremely valuable tool for understanding the complex behavior of the turbulent outer layers of stars. Some stars, such as our sun, have relatively quiet surfaces except for an occasional period of flaring; even in the most intense flares the total output of the flare is only a fraction of a percent of the total solar luminosity. In addition, a "solar wind" of gas escaping from the sun carries away about one-millionth of one percent of the sun's mass every million years. Many stars show similar behavior. Others are more extreme. Some are in an almost continual state of flaring that accounts for an appreciable fraction of their energy output. Still others are losing mass at rates of up to a few percent per million years. This wide variety of surface behavior presents a challenge and an opportunity in our efforts to understand in detail the nature of stars. Everything we know about stars we must deduce from the radiation we observe from their surfaces. Astrophysicists must deduce conditions beneath these surfaces from theories; they work out the predictions of these theories for model stars and then compare them with real stars. If these do not match, then they change the assumed characteristics of the model star, for example the central temperature and density or the abundances of the various elements, or they modify the theory, until a good fit is obtained. This process had led to a good general understanding of what the inside of a star must be like, and of how stars shine, but any detailed understanding of even the surface layers of a star is still in the future.

The sun and other stars are balls of hot gas held together by gravity. The temperature inside a star varies from millions of degrees Celsius in the center down to thousands of degrees at the surface. The visible surface of the sun, called the photosphere, has a temperature of about 6000 degrees C. Immediately above the photosphere is the chromosphere ("color sphere"), which appears bright pink when viewed just as the sun is going into or coming out of eclipse. The temperature in the chromosphere changes rapidly with height, from about 4000 degrees C just above the photosphere to several tens of thousands of degrees higher up. Above the chromosphere, the temperature undergoes a dramatic transition and increases rapidly to slightly over a million degrees in the corona ("crown"), the outermost layer of the solar atmosphere.

Because of the high temperature of the solar corona, most of its radiation is emitted at x-ray wavelengths. The study of the x-ray emission from the sun, and by analogy from other stars, is a study of the coronas of these stars. The immediate goal of this research is to understand why a star has a hot corona. Why does the temperature stop decreasing with height above the surface of a star and suddenly increase to a million degrees or more? The answer to this question turns out to involve what is happening just beneath the surface of the star. X-ray observations of normal stars indirectly give us a means of exploring conditions beneath the surface of a star.

Near the center of a star, the temperature rises to ten million degrees or more and the densities exceed that of lead. Under these conditions, nuclear fusion reactions occur. In these reactions hy-

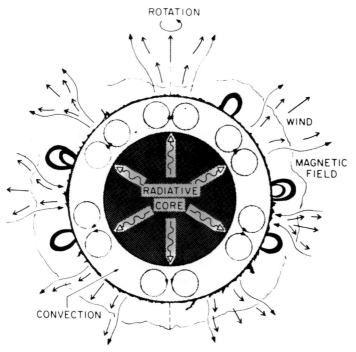

ROTATION

WIND

MAGNETIC
FIELD

RADIATIVE
CORE

CONVECTION

A schematic illustrating the structure of the sun. (R. Rosner)

drogen nuclei are fused together to form helium nuclei. Energy released in the process gradually diffuses outward to the surface of the star, where it appears as starlight.

In the first part of its journey from the center of a star, the energy diffuses outward in the form of photons, or radiant energy. Near the center of the star the photons are gamma rays of very high energy. As they move outward through the solar material, the photons are absorbed and reradiated at progressively lower-energies, from gamma rays to x-rays to ultraviolet photons and eventually to visible light. Since the total amount of energy flowing upward is the same and the photons carry less and less energy, there are more and more of the lower-energy photons in the flow. More x-rays than gamma rays, more ultraviolet photons than x-rays, and more visible-light photons than ultraviolet photons. In the sun, energy cannot be carried outward steadily and efficiently by radiation beyond a certain radius, which is about two-thirds of the way from the center to the surface. Instead, a rolling boil sets in, and the energy is carried upward by mass motions, or convection. The turbulent, boiling region is called the convection zone. The turbulent motions in the convection zone generate sound waves; the energy carried by these waves is converted to heat somewhere in the upper atmosphere of the star. For years scientists believed that this so-called acoustical heating process was responsible for producing the hot corona around the sun, and presumably around other stars as well.

Then in the early 1970s x-ray telescope observations of the sun, culminating in the Apollo telescope mount on *Skylab*, showed that the x-ray emission from the sun was highly structured; most of the hot gas was confined to magnetic loops extending high above the solar surface. Moreover, detailed examination of the motion of matter in the chromosphere with ultraviolet telescopes on *OSO-8* failed to detect the waves that were assumed to transport the acoustic energy.

All of this suggested to some scientists that the surface magnetic field of the sun plays a direct role in the heating of the solar corona, and that acoustical heating is not particularly important. They suggested instead that the conversion of energy contained in a twisted magnetic field was responsible for heating. The basis for this idea is that energy can be stored in a magnetic field if it is twisted, in much the same way that twisting a rubber band or a coiled spring increases the tension and the energy stored in these systems. The twisting also induces electric currents in the magnetized loops, much as the rotating magnetic field in the alternator of an automobile drives an electric current to charge the battery. These electric currents heat the matter in the magnetic loops. This heat appears in the form of magnetically confined loops of ultra-hot gas, which radiate x-rays. Extensive observational evidence, mostly from *Skylab*, as well as theoretical arguments, supported this point of view. Nevertheless, the acoustical heating theory had many adherents, and theories of the expected x-ray emission from normal stars continued to be based on the supposition that acoustical heating was what produced hot coronas.

The x-ray observations from *Einstein* have completed the overthrow of the acoustical heating model. The analysis by Giuseppe Vaiana, Robert Rosner, and Leon Golub of Harvard-Smithsonian and their colleagues, and by other groups, of *Einstein* observations of stars has shown that there is a strong link between x-ray emission and the average magnetic field on the surface of a star, rather than

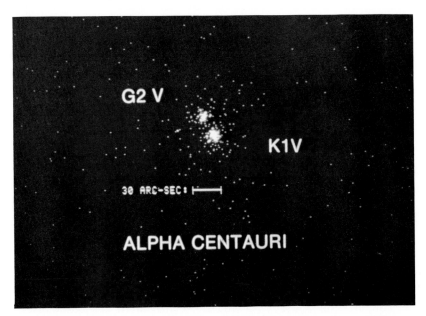

A high-resolution x-ray image of the nearby binary star system Alpha Centauri. The acoustical heating theory predicted that the star labeled "G" would be the strongest x-ray source, but the opposite is observed. (G. Vaiana, Harvard-Smithsonian Center for Astrophysics)

between x-ray emission and the predicted acoustical heating. In the magnetic heating model, the emission of x-rays from stellar coronas is determined by the strength of the magnetic field on the surface of the star and the rate at which it is twisted. The twisting of the magnetic field can be related to the interaction of the subsurface boiling of the star with the rotation of the star. The interaction of the boiling motion and the star's rotational motion with the magnetic field in the convection zone produces magnetic bubbles that float to the surface of the star, forming twisted magnetic loops.

One implication of these ideas is that stars that are rapidly rotating or have vigorous convection zones or both should have strong, twisted surface magnetic fields, which should in turn produce hot, x-ray-emitting coronas. This general scheme appears to provide a natural explanation for many of the observed stellar coronas. For example, the surfaces of a class of peculiar dwarf stars called flare stars are apparently covered with huge "starspots." These starspots are analogous to sunspots, but much larger; like sunspots, they are highly magnetized regions that are the site of large flares, or sudden unpredictable increases in the radio, optical, and x-ray emission. These flares must originate in the sudden breakdown of a highly twisted magnetic field in which large electric fields are created and magnetic energy is explosively converted to energetic particles and radiation. On the sun, flares can occur at any time, but are concentrated in cycles of about eleven years. On flare stars, flares occur almost continually and are much more powerful than solar flares.

The red dwarf flare stars have very small masses, only about a tenth that of the sun. Because of their small masses, their interiors are markedly different from that of the sun. They are in a turbulent, convective state almost throughout. Therefore, magnetic bubbles originate deep inside these stars and are amplified many times as they rise to the surface, where they appear as enormous starspots. The same process is probably at work in very young stars that are still in a state of slow collapse. These stars, which emit strong x-rays from their coronas, have yet to be stabilized by nuclear fusion reactions, and are probably convective and turbulent throughout.

The *Einstein* observations have clearly established that if a star has a large convection zone it will have a hot, x-ray-emitting corona. Can we turn the proposition around? That is, can we say that if a star has an x-ray-emitting corona it will have a large convection zone? Not in every case. Hot young stars are not thought to have strong convection zones, yet they emit comparatively large fluxes of x-rays. These x-rays are presumably produced by the large outflow of matter from the surfaces of these stars. For stars that more nearly resemble the sun, however, x-ray observations may prove to be a valuable tool for understanding the internal structure. According to the theory, stars that have masses less than about 1.7 times the mass of the sun and surface temperatures less than about 7000 degrees C should have convection zones. This prediction is based on crude estimates of the effectiveness of turbulent mixing. The theory of turbulence is so complex that the only realistic hope for understanding the details of the interior of a star is to determine by observation which type of stars have convection zones and to work back from this knowledge to establish the parameters of the turbulence. To this end, astronomers have sought to compare theoretical models with the optical spectrum of stars in the critical range. The difficulty with this approach is that elaborate models, which have a built-in uncertainty because of our ignorance of turbulence, must be compared with detailed optical observations.

The *International Ultraviolet Explorer* satellite has been used to search for ultraviolet radiation from chromospheres and the transition layers between the chromospheres and the coronas of stars. It has found evidence for the onset of convection in stars near the predicted critical mass of 1.7 times the mass of the sun, but some of the evidence is conflicting. Searches at optical wavelengths for features characteristic of chromospheric emission have yielded similar results, but it is possible that because of the relatively high temperatures and rapid rotation of these stars, the chromospheric emissions and transition-zone emissions might not show up even if a convection zone were present.

The x-ray observations, by contrast, hold the promise of unambiguously fixing the critical stellar mass below which stars are expected to have convection zones. A detailed analysis of the x-ray observations of stars near the critical mass by members of the Harvard-Smithsonian solar and stellar x-ray group shows a precipitous drop in coronal x-ray emission for stars with masses greater than 1.7 times the solar mass. This agrees with the optical and ultraviolet results. As more observations are analyzed, we should be able to firm up these results. Doing so will provide an extremely important key to the structure of stars.

Another way to understand what goes on inside a star is through what might be termed stellar autopsy. By studying the remains of an exploded star, we can piece together the circumstances of its death. Small and medium-sized stars, that is, stars having masses roughly equal to or less than that of the sun, go quietly. When the nuclear fuel of such a star is exhausted, it collapses to become a white dwarf. Unless the star is a member of a multiple star system and a nearby companion star can dump additional matter onto its surface, thus providing a new source of fuel, the star's evolution ends at the white dwarf stage. Over the course of billions of years, the white dwarf will slowly cool and fade from sight. Our galaxy probably contains several billion invisible white dwarfs.

If the white dwarf is a member of a binary star system, it may have flashes of brilliance from time to time. If the white dwarf is sufficiently close to its companion star, that is, almost touching it, the gravitational field of the white dwarf can pull matter away from the surface of the companion star. This matter typically does not fall directly onto the white dwarf; rather, it forms a disk of gas that slowly spirals inward and accretes onto the surface of the white dwarf. The accreted material is mainly composed of hydrogen and helium gas, the raw materials for thermonuclear reactions. Add to this the high density produced by the strong gravitational forces on the surface of a white dwarf and the multimillion-degree temperatures generated as the gas falls onto the white dwarf's surface, and you have an explosive situation.

Thermonuclear explosions are thought to be responsible for novas, outbursts in which a star is observed to flare up to a brightness of a million suns for a few weeks. Novas were observed by Chinese and Japanese astronomers as much as two thousand years ago; they were referred to as "guest stars" because they were apparently not permanent members of the stellar family; rather, in the manner of inconsiderate guests, they appeared suddenly and then disappeared after a few weeks or months. When these objects were rediscovered by modern astronomers, they were given the Latin name *nova* meaning "new."

We now know that novas are not new stars at all. On the contrary, they are among the oldest stars, white dwarfs that have exhausted their fuel after shining for billions of years. The accreted matter has allowed them to come out of retirement and to shine brilliantly, if only for a few weeks. The old oriental term, guest star, is after all a better description than new star.

There is a limit to how massive a white dwarf can be. If its mass exceeds a certain critical value, about one and a half times the mass of the sun, the white dwarf configuration is no longer stable. A star that has a mass greater than the critical value when its nuclear fuel is exhausted will collapse catastrophically to form a neutron star — or, if the mass is larger still, a black hole. The collapse to a neutron star precipitates an enormous explosion. A flash of light brighter than a billion suns is produced; after reaching its peak in about a week, the brightness slowly declines over the course of several months. Such events are called supernovas. They were observed and recorded centuries ago by Chinese and Japanese astronomers, as well as by the Arabs and possibly the American Indians. In post-Renaissance Europe one of the founding fathers of modern astronomy, Tycho Brahe, became famous because of his studies of such an event, but it was not until the twentieth century that astronomers understood that supernovas are distinct from, and much more powerful than, novas. This type of outburst involves the entire star, and releases a million times more energy.

Tycho Brahe made detailed observations of a supernova that occurred in 1572, observations that changed the course of Tycho's life, and quite possibly that of modern science. A man known for his acid tongue and his silver nose, Tycho rushed through life like a river out of control. He ran roughshod over tradition and sensibilities in both his life and his work, and he made many enemies as he rearranged the cosmic landscape. He overate, drank too much, fought too much, and, fortunately for science, worked too hard. At age twenty a portion of his nose was cut off in a duel with another Danish nobleman over which combatant was the best mathematician. He then shocked his peers by marrying a peasant woman. At age twenty-six he became the most famous astronomer in the western world.

It was then, in 1572, that Tycho began his celebrated study of the supernova that appeared in the constellation of Cassiopeia. His work, which showed that the supernova was located far beyond the earth and the moon, demolished faith in the Aristotelian view that the stars were perfect and unchangeable, and the universe has never again looked quite the same to human eyes. Tycho's observations, made before the invention of the telescope, are still valuable to astronomers. With a thoroughly modern approach that was foreign to his contemporaries, he carefully calibrated his quadrants and sextants to minimize systematic errors. He was also the first astronomer to determine the amount of refraction, or bending of the light rays from the stars, caused by the atmosphere, and to correct his results for it. Tycho also kept detailed records of his observations, noting the conditions under which they were made and possible reasons for discrepancies with previous observations. In other words, he did things scientifically.

Tycho's work on the supernova of 1572 made him famous, and rightly so. Science is more than the painstaking accumulation of facts. The mark of a great scientist is the ability to recognize which facts to collect and which problems to tackle. Tycho was not the only astronomer to notice the supernova of 1572. He was not even the first — Wolfgang Schuler saw it in the predawn hours of November 6, 1572, five days before Tycho. But Tycho was the one who carefully measured its position. He established beyond doubt that it did not move relative to the other stars in Cassiopeia for the 485 days it was visible, and he systematically estimated its changes over time in color and luminosity. Initially the supernova was the brightest star in the sky, comparable in brilliance to Venus. On a clear day it could be seen at noon. By December it had declined in brightness to a magnitude equal to that of Jupiter. Over the next year it gradually faded to invisibility, as its color changed from white to yellow to red and finally back to a dull white.

More than three and a half centuries later, Walter Baade of Mount Wilson Observatory used Tycho's observations to construct a curve that described the variation over time of visual brightness of the supernova—a light curve. This curve closely resembles the light curves of other supernovas known as Type I. Because of their extraordinary brilliance, supernovas in distant galaxies can be observed with modern telescopes. Every year ten or twenty supernovas are detected by "supernova patrols" set up at a few major observatories. Type II supernovas seem to occur about ten times more frequently than Type I, and tend to be about five times fainter at their maximum brightness. The optical spectrum of Type II supernovas suggests that the explosion occurs in a massive, hydrogen-rich envelope. Type II supernovas always occur in the arms of spiral galaxies, in association with bright massive stars, gas, and dust; they are rarely if ever observed in elliptical galaxies, which have little or no gas and few bright, massive stars.

All this suggests that Type II supernovas are produced by the explosion of bright stars about ten times as massive as the sun that have formed in the last few hundred million years from gas and dust in the arms of spiral galaxies. Type II supernovas are believed to be associated with the formation of neutron stars. The Crab Nebula, with its pulsar, and all the neutron stars in x-ray binaries are therefore thought to be remnants of Type II supernovas.

Type I supernovas are rarer and more brilliant than Type II supernovas. The optical spectrum during their outburst shows little or no evidence of a hydrogen envelope. Type I supernovas occur in elliptical as well as spiral galaxies and show no preference for the arms of spiral galaxies. Apparently they are produced by an older population of stars that formed several billion years ago. The more massive a star is, the brighter it shines and the more rapidly it uses up its nuclear fuel. Since very few if any new stars have formed in elliptical galaxies over the past several hundred million years, and since stars more than about five times as massive as the sun have long since blown apart, it follows that Type I supernovas are produced by stars not more than a few times more massive than the sun. But other astronomers have argued that Type I supernovas are correlated with regions of star formation in both spiral and elliptical galaxies, and are therefore associated with relatively young, massive stars. So the nature of the stars that produce Type I supernovas remains an open question.

An intriguing idea that has gained increasing support is that Type I supernovas are produced by exploding white dwarf stars. We know that if a white dwarf accretes a small amount of matter, a relatively small thermonuclear explosion will occur, blowing off the outer layers of the star and producing a nova outburst. What if a white dwarf could accrete a large amount of matter—enough to push it over the critical mass limit for stability? Then, as a number of authors have shown, the entire white dwarf would explode, producing an event that would have the characteristics of a Type I supernova. Other scientists, however, have shown that the evolution of a massive star can lead to a supernova explosion consistent with the observations of Type I supernovas.

One line of attack on this controversy is to use x-ray observations to infer the mass ejected in a Type I supernova explosion. If the explosion is produced by a white dwarf rather than a massive star, then the ejected mass should not be greater than the critical mass for the stability of a white dwarf, or about one and a half times the mass of the sun. The mass ejected in a supernova is best measured by means of high-resolution x-ray observations. This matter, which is exploded outward at speeds of thousands of kilometers per second, plows into the surrounding interstellar gas, where it produces an expanding shell of hot gas, with temperatures of several million degrees C. This gas radiates mostly x-radiation, so x-ray observations provide the best means of performing an autopsy on the remains of an exploded star. The best subject for such an autopsy is the remnant of Tycho's supernova. We know how long ago it occurred and, thanks to Tycho, we know what type of supernova it was.

Using high-resolution x-ray images made by the *Einstein* observatory, Fred Seward and his colleagues at Harvard-Smithsonian showed that the mass ejected in the 1572 supernova was in the range of one to two solar masses. Given this range and the uncertainties in their estimate, this result cannot be taken as unequivocal proof that a white dwarf exploded to produce Tycho's supernova. It does, however, constitute strong evidence for that point of view.

Additional evidence in favor of a white dwarf explosion comes from the failure to detect a neutron star in the center of the Tycho supernova remnant. According to conventional theories of the cooling of neutron stars, a 400-year-old neutron star should still retain an appreciable amount of the heat generated in its formation, and should have a surface temperature of several million degrees. The thermal radiation from the surface of such a neutron star should be easily detectable with the *Einstein* observatory at the distance of the Tycho supernova remnant. Some supernova remnants in our galaxy that are substantially older than Tycho's show evidence of a hot neutron star; the Tycho supernova remnant does not. Either some exotic cooling mechanism has cooled the neutron star much faster than is usual, or there is no neutron star there. But even if it were cool, such a young neutron star should still be generating, by analogy with the Crab Nebula pulsar, large amounts of high-energy particles, and should produce an extensive radio, optical, and x-ray cloud similar to the one in the Crab Nebula. Some supernova remnants show evidence of such clouds; in the Tycho supernova remnant no such cloud is visible. The Tycho remnant has a distinct shell-like appearance at both radio and x-ray wavelengths. The conclusion again is that either the neutron star in the Tycho supernova remnant is an unusual one or there is no neutron star there at all. The weight of all the evidence taken together — the mass of the ejected matter, the nondetectability of thermal radiation from or particle production by the hypothetical neutron star —suggests that the Tycho supernova (and by implication all Type I supernovas) was produced by an explosion in which a white dwarf blew completely apart.

Another important discovery by the *Einstein* observatory has been the detection of heavy elements in the matter ejected from the supernova. Robert Becker and his colleagues at Goddard Space Flight Center have used data from the solid state spectrometer to show that the ejected matter is rich in silicon and sulfur. This effect could be caused by some peculiar nonequilibrium ionization conditions in the expanding shell. More likely, though, the x-ray observations are giving us a measure of elements that were ejected into space during the supernova explosion.

As the shell expands, these elements will be mixed with the interstellar gas and spread throughout the galaxy. On earth, these elements and other medium-heavy elements are commonplace. Dirt, sand, and many rocks are mostly silicon and oxygen; water is mostly oxygen; air is mostly nitrogen and oxygen. Yet in stars or interstellar space these elements are not common at all. They would not be here, and neither would we, if supernovas had not manufactured them long ago and dispersed them into the interstellar gas, where they eventually became part of a cloud that collapsed to form a star with planets around it. It is a complex and awe-inspiring chain of events that has led to our presence here. But it is by no means miraculous. Our understanding of each link grows out of a combination of theory and observation, and each link we postulate must survive the confrontation with observation, or it will be discarded and a new link sought. It is a slow, exacting process, but it is the only way to be confident of our knowledge. The *Einstein* observations of the fiery remains of the supernova studied by a fiery man are helping us to understand one of the crucial links in the chain.

13

Active Galaxies and Quasars

With the flight of the *Einstein* observatory, the x-ray sky did not change so much as come into focus. *Einstein* did not conduct a complete survey of the sky, as did *HEAO-1*, so we still do not have a sharply focused x-ray picture of our galaxy. Nevertheless, by combining the results from *HEAO-1* and *Einstein*, we can make a broad-brush picture of the x-ray sky. It has the same general structure as the picture obtained at optical wavelengths, but there are significant differences. Stars, galaxies, and clusters of galaxies show up in both views. But in the x-ray sky, the brightest objects are collapsed stars, primarily neutron stars, but including at least one black hole — objects that are invisible at optical wavelengths. By the same token, none of the 100 visually brightest stars make the list of the 100 brightest x-ray sources, although stars of almost every type are weak x-ray sources. An x-ray picture of our galaxy also prominently displays the remnants of supernova explosions, giant shells of hot gas that also show up brightly at radio wavelengths but are only faintly visible at optical wavelengths.

As we go beyond our galaxy we encounter other galaxies that are more or less like our own. The x-ray images of these galaxies look like the skeleton of the optical picture, with a few bright neutron stars and possibly black holes scattered around in roughly the same pattern as the optically bright stars. This is not surprising, since a neutron star or a black hole will not be a strong x-ray source unless it has a nearby companion star, which more often than not is an optically bright star.

In the nearby galaxies, as in our own, a bright source of x-rays has been detected at the nucleus of the galaxy. Other, more distant

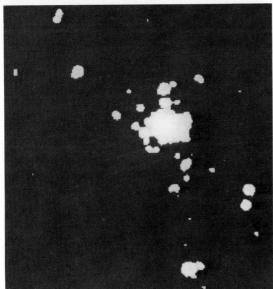

Optical (top) and x-ray (bottom) images of the Andromeda galaxy. (Optical photo, Palomar Observatory; x-ray photo, L. Van Speybroeck, Harvard-Smithsonian Center for Astrophysics)

galaxies show much more intense radiation from their nuclei. X-ray observations of more than a hundred galaxies suggest that the only difference among the nuclei of normal galaxies, active galaxies, and quasars is the power level of the central source. What is the central source? What powers it? A combination of radio, infrared, optical, and x-ray observations appears to be closing in on the answer.

The search for the central source of power in quasars parallels in many ways the search for the power source of x-ray stars. Quasars, like x-ray stars, were discovered as the result of the development of new technology in astronomy. In the late 1950s radio astronomy was undergoing the same kind of explosive development that would be experienced in x-ray astronomy a decade later. Strong radio emission was detected from supernova remnants, giant radio galaxies, and some peculiar starlike objects that were called radio stars. One of the first steps astronomers would take after the discovery of a new radio object was to find its optical counterpart. Since they were familiar with the optical appearance of the various objects in the sky, in most cases they could classify a new object imme-

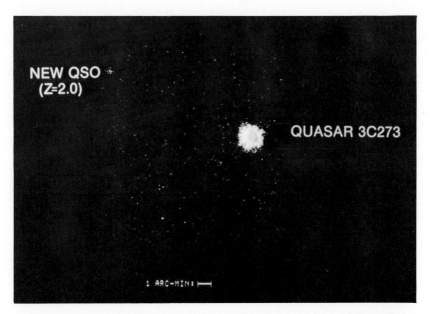

An x-ray image of 3C273. (J. P. Henry, Harvard-Smithsonian Center for Astrophysics)

diately as a supernova remnant, a galaxy, or something else. Equally important, the optical spectrum of these objects was usually rich in information about their dynamics and distance. Astronomers could tell, for example, whether the object was expanding and whether it was rotating, and by using a variety of techniques they could determine its distance from earth. This is extremely important, because knowing the distance is crucial in converting the observed brightness to intrinsic brightness.

By 1960 a number of radio stars were known, but none of them had been identified with optical counterparts. The positions of the radio stars were so ill-defined that there were simply too many possibilities. Then in 1960 Thomas Matthews, a radio astronomer at the California Institute of Technology, got an improved radio position on a source known as 3C48 (number 48 in the Third Cambridge Catalog of radio sources, the Cambridge in this case referring to Cambridge, England, where much of the pioneering work in

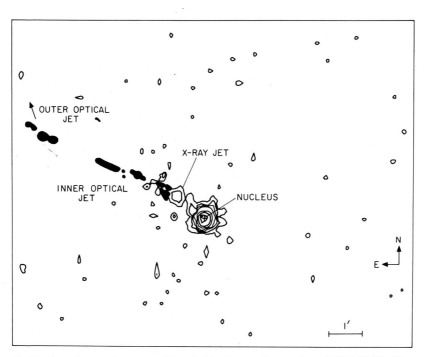

Optical and x-ray images of the jet in the active galaxy NGC 5128. (E. Schreier et al., *Astrophysical Journal (Letters), 234* (1979) L39)

radio astronomy was done). Matthews gave this information to his colleague Allan Sandage, who immediately began a search with the 200-inch telescope on Palomar Mountain. In his book *The Red Limit*, Timothy Ferris reports Sandage's recollections: "I took a spectrum that night and it was the weirdest spectrum I'd ever seen. I took the spectrograph off the telescope and put the photometer on, to check its colors, and the colors were different than any object I'd ever seen. The thing was exceedingly weird."

What was so weird about the spectrum of 3C48? For one thing, the relative amounts of red, blue, and ultraviolet light did not match any known stellar spectrum. The object called 3C48 had large excesses of blue and ultraviolet light. More perplexing were the spectral lines. Each chemical element produces its own characteristic pattern of peaks or dips in the spectrum, corresponding to its emission or absorption of radiation. It was through spectral measurement that the element helium was discovered on the sun more than a century ago, before it was discovered on earth. Over the years, as the instruments for measuring spectra and the theories about how spectra from different elements should look have improved, astronomers have become highly skilled at unscrambling complex stellar spectra. They can tell how much of each element is present in a star and estimate its temperature and density; by using the Doppler effect, they can tell whether the material is rotating, expanding, or orbiting in a binary system, and in each case they can estimate the speed of the motions. But for all their skill, the Cal Tech astronomers could not figure out the spectrum of 3C48.

As more radio stars were identified, the mystery only deepened. The name "radio star" was replaced by "quasi-stellar radio source," expressing the growing suspicion that these objects were unlike any stars ever studied before. This name was soon shortened, at the suggestion of NASA astrophysicist Hong Yee Chiu, to "quasars." The name stuck, in spite of complaints from some who argued a bit peevishly that it was misleading and ugly. On the more substantial issue of what the quasars were, astronomers continued to be perplexed. None of the spectra of the quasars could be understood in terms of known elements.

Jesse Greenstein of Cal Tech, perhaps the world's top spectra detective, came up with a complex explanation in terms of a collapsed stellar remnant of a supernova; this looked promising for a while, but eventually it was shown to be incorrect. As Ferris recounts it, Greenstein and Maarten Schmidt, another Cal Tech astronomer, once toyed with the idea that the peculiar quasar spectra could be the result of large "red shifts" in the lines. In other words, the entire pattern could be shifted to lower (that is, redder) wavelengths as a result of motion away from the observer. They dismissed the idea. "It was," reflected Greenstein, "a classic story of inhibition of one's creativity by knowing too much."

More than a year later, Schmidt came back to the idea. In February 1963, while he was studying the spectrum of the quasar 3C273, it occurred to him that the simplest explanation was that the spectral lines of 3C273 were due to a system of lines of hydrogen, the most common element, that were red-shifted by 16 percent. He rushed to Greenstein's office to tell him the news. Together they applied this idea to 3C48. A red shift of 37 percent would explain everything!

Today, the spectra of hundreds of quasars have been analyzed, and red shifts ranging from slightly less than 10 percent to 350 percent have been measured. What produces these red shifts? The simplest interpretation is that they are produced by the expansion of the universe. In the late 1920s Edwin Hubble of Cal Tech showed that the distant galaxies are moving away from us at a rate that increases with their distance. This shows up as an increase in red shift with distance. For example, a galaxy with a red shift of one-tenth of one percent is only half as far away as a galaxy with a red shift of two-tenths of a percent. If this explanation is applied to quasars, it means that 3C273 is about 2.5 billion light-years away, and that 3C48 is about 6 billion light-years away. Many of the observed quasars must be more than 10 billion light-years away.

This conclusion immediately raised the stakes in the quasar game. Quasars were easily the most distant objects we had ever observed. As such they could be used as probes of the large-scale structure of the universe, and they assumed a cosmological importance quite apart from an understanding of exactly what they were. Accordingly, a great deal of effort has gone into statistical analyses of the number of quasars in various regions of the sky, as well as the change in the number of quasars with distance. In astronomy, distance is measured in light-years, that is, the distance light travels in a year. So, as we probe further away from the earth in distance, we travel back in time. This effect is of little consequence for studying objects in our galaxy; the distances involved are typically thousands of light-years, and a thousand years is short compared to our galaxy's estimated lifetime of ten billion years. For quasars, however, the distances involved correspond to times that are comparable to the age of galaxies. The study of the variation of the number of quasars with distance should therefore tell us about one of the crucial times in the history of the universe, the time when galaxies were formed. No one knows exactly when galaxies formed, or how they formed. We only know that all relatively nearby galaxies are approximately the same age, ten billion years. If this is so, then as we look at greater and greater distances, that is, back to earlier and earlier times, we should eventually reach back to a time when the universe was so young that galaxies did not yet exist. The difficulty with this approach is that galaxies are not bright enough to be seen at distances of billions of light-years.

But the quasars are. Schmidt was the first to show that the number of bright quasars has changed with time; that bright quasars were far more numerous in the past. A recent survey by Schmidt and Richard Green of the University of Arizona and Herman Marshall at the Space Telescope Science Institute and his colleagues indicates that quasars were very common when the universe was one or two billion years old. At longer distances, corresponding to ages earlier than one or two billion years, no quasars have been found. This could be an indication of the conditions in intergalactic space at that time. Perhaps space was filled with dust and gas that absorbed the quasar radiation, so it could never reach us. Or perhaps the immediate environment of the quasars was cluttered with dust and gas that absorbed their radiation. Or perhaps there were no quasars. The search for answers to these questions leads very quickly into the deep waters of cosmology, which we will avoid for now. Instead we will explore the meandering currents of opinion about what might be the energy source for the quasars.

A bright quasar produces as much energy as a thousand galaxies combined. Yet the starlike appearance of quasars suggests that they must be much smaller than galaxies. This led many astrophysicists to suspect that quasars were somehow associated with the nuclei of galaxies. In 1963 Geoffrey and Margaret Burbidge of the University of California at San Diego and Allan Sandage published an important paper summarizing the evidence for the occurrence of violent events in the nuclei of galaxies; they suggested that these events were small-scale models for the quasar phenomena. Since then, radio, infrared, and optical observations have confirmed this idea. There appears to be continuum of explosive activity in the nuclei of galaxies, ranging from the relatively weak but by no means quiescent nuclei of normal galaxies such as our own up to quasars, of which the nuclei are apparently so bright that they outshine by a thousand times the galaxies that harbor them.

We observe the nucleus of a galaxy as a small region of high luminosity in the center of the galaxy. Bright nuclei appear on optical photographs of most galaxies. It is impossible to see the nucleus of our own galaxy with optical telescopes because of the absorption of radiation by intervening gas and dust; observations of nearby galaxies indicate a dense concentration of stars within a region having a radius of a hundred light-years or less. About the same time that the Burbidges and Sandage published their empirical study of violent activity in galactic nuclei, Fred Hoyle of Cambridge University and William Fowler of Cal Tech proposed that the energy source for quasars and active galactic nuclei was a superstar with a mass equivalent to about a million suns. Most astronomers agreed that a large concentration of mass was necessary to explain the enormous energy output of quasars, but they expressed doubts as to the existence of supermassive stars. Thomas Gold of Cornell suggested instead that the mass concentration was in the form of a dense cluster of stars; occasionally two stars would collide, releasing tremendous amounts of energy. Stirling Colgate of the New Mexico Institute of Mining and Technology modified this model, arguing that as a rule stars would not release large amounts of energy upon colliding; instead they would stick together to form progressively larger stars that would rush through their evolution and undergo a supernova explosion. The power output for quasars, Colgate suggested, came from a dense cluster of stars in which about ten to a hundred supernovas were occurring per year. Edwin Salpeter of Cornell suggested that quasars contained supermassive black holes, and that the quasar was fueled by accretion of matter onto a black hole. None of these ideas acquired anything approaching a majority of support among astrophysicists; the most talked-about model was probably Colgate's, because he had worked it out in most detail.

When pulsars were discovered in 1968 and a rotating neutron star was found to explain the energy output of the Crab Nebula supernova remnant, a new class of models sprang to the minds of the theorists. If a rapidly rotating object explained the heretofore enigmatic Crab Nebula, then why not the quasars? It was suggested that a large collection of neutron stars in a dense star cluster was the power source for quasars. Even more inventive was the idea that a supermassive, rotating, magnetized cloud, a giant pulsar, or "spinar," was the power source. Then, when the *Uhuru* observations of Cygnus X-1 identified Cygnus X-1 as a black hole, black-hole models experienced a renaissance, mostly through the work of Martin Rees, Donald Lynden-Bell, and their colleagues at Cambridge University.

In the meantime, improved observational techniques were having their impact on the question of what powered the quasars. Of particular importance was the detailed picture of galactic nuclei that began to emerge. In his book *The Milky Way*, Bart Bok gives a detailed description of the nuclear region of our galaxy. It appears as a bright infrared source. Infrared observations imply that at least two million stars are concentrated within a radius of three light-years. For comparison, there are *no* other stars within three light-years of the sun. In this same region infrared and radio observations have detected large clouds of gas. These clouds appear to be orbiting rapidly around a supermassive core that has the mass of five million suns. Radio observations indicate that this core is about the size of our solar system.

Observations of other galaxies suggest much of the same picture. Over the past decade, radio observations have been particularly important in establishing the nature of explosive activity in galactic nuclei and quasars. They have provided strong evidence that galactic nuclei and quasars explode repetitively, ejecting clouds of high-energy particles, always along the same axis of symmetry. When these jets of particles are large enough, they show up at optical and x-ray wavelengths. The giant elliptical galaxy M87 has such a jet projecting from its nucleus, as does 3C273. In almost every case, the radio jets extend out from the nucleus and point toward much larger bubbles or lobes of high-energy particles. In some cases these alignments stretch over tens of thousands and even hundreds of thousands of light-years. The alignment of the jets with the large lobes implies that the explosive ejection of high-energy particles has been channeled in the same direction with a precision of a few percent over a time span of several million years.

Such a well-defined axis of symmetry strongly suggests a single supermassive object. But what type of supermassive object? The conventional theory of stellar structure, applied to a star with the mass of a million or more suns, showed that such a star would collapse to become a black hole in a few thousand years. Attempts to save these stars by assuming that they rotate rapidly or that they are supported by turbulent motions seem to have failed. Gravitational

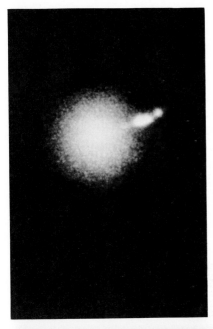

Optical (top) and x-ray (bottom) images of the active galaxy M87; note the jet-like structure emanating from the nucleus. (Optical image, Lick Observatory; x-ray image, P. Gorenstein and E. Schreier, Harvard-Smithsonian Center for Astrophysics)

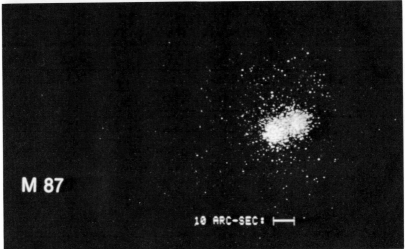

collapse seems to be the inevitable final state of such massive objects. This implies that the sources of activity in quasars and in the nuclei of galaxies are black holes.

The case for black holes as the power generators for quasars and galactic nuclei is closely analogous to the case for believing Cygnus X-1 is a black hole. It rests on three main points: (1) there is observational evidence of the existence of dark, massive, compact objects in quasars and galactic nuclei; (2) there is evidence of the release of large amounts of energy in a small region of space; (3) a black hole is the only massive object known to theory that is stable, compact, and capable of producing the observed emission.

In the case of Cygnus X-1, the evidence of the presence of a massive object came from observations of the orbital motions of the optical counterpart. For galactic nuclei we have similar evidence. The orbital motions of clouds in the nucleus of our galaxy imply a concentration there of a dark matter with the mass of five million suns, and radio observations suggest that this mass is confined to a very small region. Further evidence is provided by radio observations made with the Very Large Array of radio telescopes near Socorro, New Mexico, by Kwok-Yung Lo and Mark Claussen of Cal Tech. They observed streams of matter converging on the center of the galaxy. The observed velocities of the gas in the streams indicate that the matter is spiraling into a dark, massive object in the center of the galaxy. Even more exciting are the observations of the nucleus of the giant elliptical galaxy M87. Two groups of astronomers, working with the 158-inch telescope on Kitt Peak and the 200-inch telescope on Palomar Mountain, have found that the average velocity of the stars in the nucleus of M87 rises rapidly toward the center of the nucleus. This is the opposite of what would be expected if all the mass were in the form of stars. The observations imply a black hole with a mass about 500 million times the mass of the sun.

In Cygnus X-1, the observed variability of the x-ray emission provides evidence of the release of a large amount of energy in a small volume of space. The situation for quasars and active galaxies is analogous. The *Einstein* observatory has detected x-ray emission from more than a hundred quasars, at an intensity comparable to that observed at optical wavelengths. When these data are combined with *HEAO-1* observations at much higher x-ray energies, it becomes clear that an appreciable fraction of the energy radiated by quasars comes out as x-radiation. In at least six cases, intense x-ray flares that last about a day have been observed. This duration limits the size of the x-ray-emitting region to a volume of space with a diameter less than the distance light can travel in a day, that is, to a region roughly the size of the solar system. This is consistent with what we would expect if the gravitational energy-conversion process responsible for the activity in the nucleus were occurring near the gravitational radius of a black hole with a mass about a billion times the mass of the sun.

We have discussed how the instability of supermassive objects makes black holes seem to be the only possible power sources for galactic nuclei and quasars. However, in the words of James Gunn of Cal Tech, it is not enough to have a "monster," that is, a black hole. You must have food for the monster in the form of infalling gas. In quasars, every year the black hole must accrete a gaseous mass equivalent to the mass of one or more suns. In Cygnus X-1 the gas is supplied at a much lower rate by matter pulled off the companion star by the strong gravitational field of the black hole. In a galactic nucleus there appear to be two good possibilities: either the gas could be supplied by the millions of stars that are in orbit around the black hole, or it could be gas that falls into the galaxy from outside.

The stars in orbit around the black hole are concentrated into a region only a few light-years across. Occasionally such stars collide with one another at high speeds and are torn apart. The resulting stellar debris is then pulled in toward the black hole. Some of the collisions are gentle, and the stars coalesce to form larger stars that then run through their evolution and explode as supernovas, ejecting still more gas into the maelstrom. Still other stars may be pulled apart by the tidal forces of the black hole.

The gas to feed a black hole need not come from the galaxy. If a galaxy is embedded in a cluster of galaxies, at the cluster's center of mass, then matter lost by other galaxies in the cluster may fall into the central galaxy and find its way to the nucleus and into the black hole. Or a collision between two galaxies may result in the transfer of a large amount of gas from one galaxy to the other. It is intriguing that many active galaxies are either at the center of clusters of galaxies or have recently collided with other galaxies. Quasars are so far away from earth that any companion galaxies are difficult to detect, so it is impossible to know if this effect is important for them; future observations with NASA's Space Telescope, scheduled for launch in June 1986, may tell us.

However the gas is produced, it falls into the black hole, probably as part of a gaseous disk. As the matter gets closer to the black hole, it swirls faster; frictional processes may heat the disk to 100,000 degrees C. This high temperature may be the source of much of the optical radiation from quasars. Near the inner edge of the disk, not far from the gravitational radius of the black hole, the heat is so intense that the disk may break up into clumps of matter, which are heated to very high temperatures, producing x-radiation. Recent infrared, optical, and ultraviolet observations of the quasar 3C273 by Matthew Malkan and Wallace Sargent of Cal Tech show that its spectrum is consistent with an accretion disk, as suggested earlier by Gregory Shields of the University of Texas.

In most quasars, the radiation from the infalling matter should be intense, especially near the inner edge of the disk. If the pressure of this radiation exceeds a certain critical value, the inner edge of the disk will bloat up and form steep walls around the black hole. Inside the disk, radiation pressure and electromagnetic fields generated in the turbulent gas may lead to huge flares and the ejection of gas perpendicular to the plane of the disk. This ejection will occur very near the black hole, where the radiation pressure and the magnetic fields are strongest, and the steep wall of the disk may collimate the gas into narrow jets. These jets may be maintained for as long as the rate of the inflow of mass exceeds a certain critical value, which will depend on the mass of the black hole; over the course of millions of years, extensive jet structures, similar to the observed ones, might be produced.

If there is no outside supply of gas, the galactic nucleus will settle down to a relatively quiet existence once most of the stars around it have been swallowed by the black hole or disrupted and most of the gas has been swallowed by the black hole or ejected in a jet. This may be the situation in the nucleus of our galaxy. Every hundred million years or so, one of these quiet galactic nuclei may flare up when a globular star cluster or a large gas cloud spirals into the nucleus. Episodes such as this may be what we are observing in some of the relatively nearby active galaxies.

This black-hole model of galactic nuclei, which incorporates some of the features from most of the earlier models for quasars, also explains why most galactic nuclei are active at some level or other. They all presumably contain massive black holes. Some nuclei have large black holes and large supplies of gas, others have small black holes and small supplies of gas, but the model in its simplest form suggests that every nucleus was once a quasar and that every quasar will eventually become the nucleus of a normal galaxy. Normally, a quasar might use up its gas supply and turn into a normal galaxy in a few million years. Galaxies appear to be about ten billion years old. If the quasar phase lasts only a few million years, or a few ten-thousandths of the lifetime of a galaxy, then as we search the universe we should find only a few ten-thousandths as many quasars as normal galaxies. And this is about what we do find.

Have we solved the enigma of the quasars? Possibly. The outlines of the model seem to provide a coherent structure for understanding what we observe. But the structure is supported in many places by only a few observations, or by theory that is more speculation than substance. In the next decade, observations with the Space Telescope, the Very Large Array of radio telescopes, the planned Advanced X-ray Astrophysics Facility, as well as other radio, infrared, optical, and x-ray observatories, coupled with theoretical work, will decide whether the black-hole model will stand.

14

Clusters of Galaxies and the Missing Mass

X-ray observations of galactic nuclei and quasars allow us to indicate the presence of matter near a black hole — a black hole being the state of ultimate collapse. In clusters of galaxies, x-ray observations reveal the opposite extreme. The hot gas detected between the galaxies in a cluster is very thinly spread: so thinly that the same number of particles that are present in one quart bottle of air at atmospheric pressure are dispersed over a volume as large as the earth. Yet this gas showed up clearly on the *Einstein* observatory's x-ray detectors, because of its enormous mass. Clouds of hot gas millions of light-years across are routinely detected in association with clusters of galaxies. These clouds are a dominant feature of the x-ray sky, yet they are completely invisible to optical, infrared, and radio telescopes. Studies of the x-radiation from clusters of galaxies has proven to be a powerful method for attacking problems relating to both galaxies and clusters of galaxies.

One of the most fundamental of these problems is measuring the mass of galaxies and clusters of galaxies. By observing the orbital motions of stars and clouds of gas in a galaxy, we can compute what the galaxy's mass must be to keep the stars and gas clouds in orbit. This mass, which we call the gravitational mass, is several hundred percent larger than the observable mass, that is, the mass that can be observed with a telescope of any type, be it a radio, optical, or x-ray telescope. Fritz Zwicky of Cal Tech first noticed this problem in the 1930s. He came across it while studying the motions of galaxies in clusters of galaxies, and he gave it a catchy name: the problem of the missing mass. Similar discrepancies between the observed mass and the gravitational mass were noted by Jan Oort of the University of

Leiden in 1941 and by Martin Schwarzchild of Princeton in 1954, but otherwise most astronomers seem to have ignored the problem until the 1970s. The feeling was that the problem would probably go away as observational techniques improved. Either the apparent discrepancy would be shown to be spurious or the missing mass would be found in some other form, such as hot gas.

As x-ray techniques improved, this possible escape route was eliminated. Hot gas was detected in clusters of galaxies; the mass of this gas was large, comparable to the observed mass in galaxies, but it was not nearly large enough to account for the discrepancy between the gravitational and the observed mass. The missing-mass problem was raised anew in 1974 by Jeremiah Ostriker and his colleagues at Princeton and by Jan Einasto and his colleagues at Tartu Observatory in Estonia. Their publications created a flurry of interest and some mild clashing of swords by opposing groups at meetings, but most astronomers stayed on the sidelines. The data were still not good enough.

That situation changed in the late 1970s with the development of highly sensitive devices for detecting optical radiation and of much more sensitive radio and x-ray telescopes. Vera Rubin and her colleagues at the Carnegie Institution began using these greatly improved detectors to study the dynamics of spiral galaxies. They were able to demonstrate beyond doubt that the problem of missing mass is a real one for spiral galaxies. Using the Doppler effect to measure the speed at which stars or clouds of gas orbit the center of a galaxy makes it possible to determine the mass inside the orbit. Optical photographs of galaxies show that the light from a galaxy falls off steadily from the central to the outer regions until it eventually fades out into the background. There are not as many bright stars or gas clouds on the outer edges of galaxies as farther toward the center, so it is difficult to make the crucial measurements. In the past it would take hours to determine orbital speeds for even the bright inner regions of most galaxies. By contrast, with modern techniques an accurate determination of velocity at large distances from the center of a galaxy is possible in about three hours of observing time. In this way Rubin and her colleagues were able to obtain accurate estimates for the masses of the galaxies. The mass of a galaxy can be estimated independently by assuming that the optical image is produced by a collection of normal stars similar to those in our galaxy. This is a good assumption, because the spectrum of light from spiral galaxies is consistent with it. What Rubin and her colleagues found is that the mass calculated from the orbital motions is much greater than the mass deduced from the optical image of the galaxy. Evidently the galaxies are immersed in an envelope of invisible matter.

Radio observations have confirmed and extended these results. They show that the invisible envelopes of spiral galaxies must contain about ten times as much mass as is apparent from the radio and optical observations. One of the most dramatic examples of an invisible envelope around a galaxy comes from x-ray observations of the giant elliptical galaxy M87. M87 is in the center of a rich cluster of several thousand galaxies in the constellation of Virgo. It is a bright, well-studied galaxy that has played a key role in many of the exciting developments of the past two decades. The bright jet protruding from its nucleus was the first of many similar nuclear jets that were discovered in quasars and galactic nuclei. The strongest evidence of the presence of a black hole having the mass of more

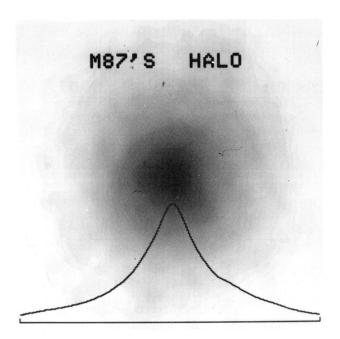

An x-ray image of the extended halo of the giant elliptical galaxy M87. (W. Forman and C. Jones, Harvard-Smithsonian Center for Astrophysics)

than a billion suns comes from optical studies of the nucleus of M87. It is one of the largest and brightest galaxies in the Virgo cluster. In the late 1960s, Halton Arp of Cal Tech and Francesco Bertola of Italy's Padova Astronomical Observatory reported a large, faintly visible envelope around M87. This suggested that the mass of M87 was at least ten times larger than that of our galaxy. In the 1970s, x-ray observations indicated that M87 was enveloped in a large cloud of hot gas. The *Einstein* x-ray observatory made it possible, through detailed observations of this hot corona, to study the mass and extent of the invisible envelope around M87.

Using *Einstein* data, Daniel Fabricant and Paul Gorenstein of Harvard-Smithsonian were able to trace the density, temperature, and pressure in the corona out to distances of several hundred thousand light-years. These observations indicate that the pressure of the gas is balanced by the gravity of M87. Therefore, measuring the distribution of pressure makes it possible to measure the gravitational forces and, by implication, the mass of M87. Fabricant and Gorenstein found that M87 has the mass of thirty trillion suns, 90 percent of which must be hidden in some so-far-unobservable form. This is thirty times the mass of our galaxy, even taking into account the invisible massive envelope of our galaxy. Are all giant elliptical galaxies so massive, or is M87 special?

X-ray observations of other giant elliptical galaxies in the Virgo cluster indicate that M87 is indeed special. The hot coronas around these other galaxies show no similarity to that of M87. The difference is probably due to M87's special position at the center of the cluster. Whereas most of the galaxies in the cluster are moving rapidly back and forth in the cluster, like a swarm of bees around a beehive, M87 appears to be sitting motionless, like the beehive, at the center of the cluster. Apparently it is at the cluster's gravitational center, or center of mass. This position might explain its large mass. Galaxies not at the center of mass may have their envelopes pulled away by collisions with one another or by the gravitational field of the cluster as a whole. These detached envelopes drift to the center of the cluster, where they are assimilated by M87. Thus, over the course of ten billion years, M87 has grown at the expense of other galaxies.

This sequence agrees with x-ray observations of other clusters. Wherever a galaxy is found at the center of a rich cluster of galaxies, it has a large x-ray corona and, by implication, a large envelope of invisible mass. Rich clusters without centrally located galaxies are also bright x-ray sources. Instead of being centered on any one galaxy, the gas in these clusters is symmetrically distributed around the center of mass of the cluster. It is possible to use the same techniques as for M87 to measure the gravitational mass of a cluster of galaxies. Once again, the gravitational mass is found to be about ten times larger than the mass that can be measured in the hot gas through x-ray observations or in the galaxies through optical or radio observations. These results are consistent with optical observations, which show that the motions of galaxies orbiting around the cluster's center of mass imply a gravitational mass about ten times larger than the observed mass of the stars and gas in the galaxies.

The discrepancy between gravitational and observed mass is not observed in the computation of the motions of planets in the solar system or stars in a binary system. It seems to be a problem only on a galactic scale or larger, where systems with diameters of hundreds of thousands and millions of light-years are involved. Two general explanations are offered for this. One is that the theory used to calculate the gravitational mass is wrong. The other is that the theory is right and the mass is hidden in some form that has so far escaped detection.

Mordehai Milgrom of Tel Aviv University has considered the possibility that the theory is wrong, and has suggested a modification of the theory that might solve the missing-mass problem. His efforts have drawn responses ranging from "crazy" to "interesting," but have generated little enthusiasm so far. The possibility that the theory of gravitation and dynamics, also known as Einstein's General Theory of Relativity, is wrong is just not appealing to most astrophysicists. Einstein's theory has tremendous aesthetic appeal because of the simplicity of its assumptions, and it has worked extremely well up to now, passing every experimental test with flying colors. Nevertheless, this does not guarantee that the theory is right. After all, Newton's theory of gravitation is a completely adequate theory for most situations, but it does not explain everything: for example, black holes. The same could be true of Einstein's theory. But why go to the trouble to change a theory that is working well in most cases before you have exhausted all reasonable alternative explanations? Such practical conservatism characterizes the attitude of most scientists; the primary effort to explain the missing mass has gone into searching for forms in which most of the mass in the universe could be hidden.

It cannot be hidden in the form of gas. Cool gas would show up because of its absorbing properties. Hot gas would show up because of the radiation it would produce. There is evidence of the presence of both cool gas and hot gas in galaxies and clusters of galaxies, but the amount of gas implied by the observations is not nearly enough to explain the missing mass. Dust grains can also be ruled out. If clusters of galaxies were filled with enough dust to account for the missing mass, then dust would absorb the blue light from the galaxies in the cluster. The galaxies would take on a distinctively red glow, like the sunset on days when the atmosphere is filled with smoke, smog, or dust. In addition, the energy of the absorbed radiation would be reradiated at infrared wavelengths. If the missing mass were in the form of dust, then galaxies and clusters of galaxies should be embedded in bright infrared envelopes. The failure to observe either the reddening of galaxies or infrared envelopes around galaxies rules out dust as the form of the missing mass. Yet another argument against dust is that it is made of medium-heavy elements such as carbon, silicon, and iron. If 80 or 90 percent of the mass of the universe were in the form of dust grains, then most of the matter in the universe would be in the form of medium-heavy elements. This is completely contrary to the generally accepted theory, which is strongly supported by observational evidence, that 99 percent of the atoms in the universe are in the form of hydrogen and helium.

Suppose instead that the missing mass is hidden in the form of stars of very low mass. The stars of very low mass, red dwarfs with masses about 10 percent that of the sun, have not yet evolved to the white dwarf stage. If 90 percent of the matter in our galaxy is in the form of low-mass stars, then there should be far more red dwarf stars than there appear to be. It is possible that the red dwarfs might be concentrated in the outer parts of the galaxy, where they might have escaped detection so far. As discussed in Chapter 12, red dwarf stars produce intense radio, optical, and x-ray flares. This property should make it possible for advanced x-ray observatories, working together with radio telescopes and the Space Telescope, to refine the estimates of the number of red dwarfs in our galaxy, and decide whether or not they are the solution to the missing-mass problem.

It will be much more difficult to rule out brown and black dwarf stars. Brown and black dwarfs, which are essentially freely wandering Jupiter-like objects, may be so dim that it is impossible ever to detect them. Though there is no sound theoretical reason to believe that they exist in the required numbers, it is possible that such objects were produced in large numbers long ago, when galaxies were just beginning to form. Because of their low mass, they would then have formed a large halo around their galaxy. Even if there were quadrillions of them around every galaxy, they might escape detection.

At the other extreme is the suggestion that most of the mass has been locked up in black holes. If these objects were formed from stars having masses of thirty or so times that of the sun, then the galaxies would have been exceedingly bright before these stars turned into black holes. Since no such bright galaxies have been detected (the quasar radiation comes from a very small volume of space; these galaxies, on the contrary, would be huge), the era of formation of black holes has to be pushed back into the distant past, when the universe was only a few hundred million years old, only a tiny percentage of its present age. Yet another possibility is that early in the history of the universe most of the matter collapsed very quickly into supermassive black holes that are swarming around the outskirts of galaxies. Advanced x-ray facilities may be able to test this hypothesis. Supermassive black holes may produce detectable variations in the x-ray emission from the coronas of galaxies such as M87, either through their effect on the distribution of the gas, or through the radiation produced by accretion of gas into the black holes.

Cosmological considerations argue against the missing mass being in any normal, baryonic form. By baryonic we mean the normal constituents of the atomic nucleus, protons and neutrons, as opposed to other particles of the type we are about to discuss. According to the Big Bang model, the universe as we know it began expanding from a hot dense state about fifteen billion years ago. In the first few minutes of this expansion the deuterium (heavy hydrogen) in the universe was created. The amount of deuterium created depends very sensitively on the conditions during this early stage. If the mass density was too high, the deuterium would all have been processed into helium. By observing the amount of deuterium in the universe we can set limits on the mass density of normal matter. The observations suggest that the concentration of deuterium is barely consistent with most of the missing mass being in the form of normal matter. The observations of deuterium are uncertain, however, by about 50 percent. If more accurate future estimates indicate a higher concentration of deuterium, then the missing mass must be in some nonbaryonic form.

One such form is the neutrino. Neutrinos are subatomic particles that are produced in certain nuclear reactions. Nuclear reactions of this type are thought to have been so common in the early universe that the universe should be full of neutrinos. Since neutrinos were thought to have no mass, however, and to interact only very weakly with matter, they did not seem important for the question of the missing mass. Then in 1980, experiments in the United States and the Soviet Union suggested that neutrinos might have small masses. Even a very small mass could make a very large difference, because there are thought to be at least a hundred million times more neutrinos than protons in the universe. The mass indicated by the experiments was about one-ten-millionth of the mass of the proton, enough to explain the missing mass. Has the solution to the mystery of the missing mass been found, not in some remote galaxy, but in an earthbound laboratory?

Once again, the answer is not clear. Three problems have arisen. First, further experiments have clouded the issue of whether neutrinos really have mass, and if so how much. Secondly, it has proven difficult to understand how it is possible to form galaxies in a universe dominated by fast-moving neutrinos. Theory suggests that clumps much larger than clusters of galaxies would form first, then clumps the size of clusters of galaxies; finally, clumps the size of galaxies would condense from these "superclusters." Yet a variety of observations indicate that the formation happened in the opposite order.

For example, x-ray images of clusters of galaxies show a variety of types. As discussed by William Forman and Christine Jones of Harvard-Smithsonian, the various types can be arranged into three groups: early, intermediate, and evolved. The early systems have a low x-ray luminosity, a low central-galaxy density, and an irregular x-ray shape consistent with the irregular distribution of galaxies. In the intermediate systems, the structure has begun to evolve toward a more symmetric shape: instead of many separate concentrations of x-ray emission there are just two large x-ray clumps of gas and galaxies. In the evolved systems, the cluster has a high x-ray luminosity, a high central-galaxy density, and a symmetrical structure. This sequence of types can be readily understood in terms of a swarm of randomly moving galaxies coming to equilibrium with

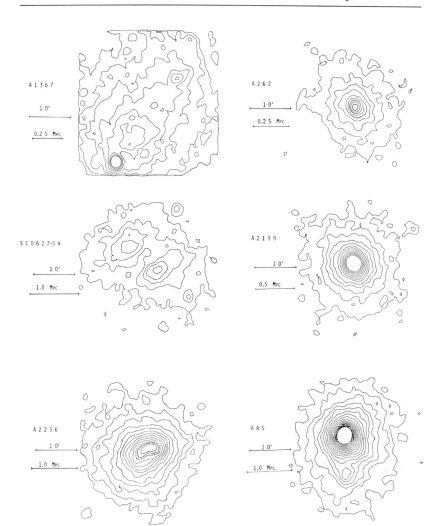

X-ray maps of clusters of galaxies, indicating the classification into early (top), intermediate (middle), and evolved (bottom) systems. The clusters on the right have a centrally located supergiant galaxy; those on the left do not. (W. Forman and C. Jones, Harvard-Smithsonian Center for Astrophysics)

each other, and it strongly suggests that the galaxies formed before the clusters. Because of different initial conditions, different clusters would be expected to be in different stages of their evolution toward equilibrium. The next generation of x-ray observatories will have the exciting potential of tracing the evolution of clusters back to early stages, perhaps to the early epochs of the universe when clusters and galaxies were formed. We should then have direct evidence bearing on the question of whether galaxies were formed first, and then clustered together, or whether the clusters were the matrix from which the galaxies condensed. For now, the preliminary evidence is against the idea that neutrinos make up the missing mass.

The neutrino hypothesis also suggests that the fraction of missing mass around galaxies should be much less than in clusters; the observations disagree with this prediction. The difficulties encountered by the neutrino hypothesis have prompted astrophysicists to consider other exotic particles that could have been formed in the Big Bang. One of these is the axion. Unlike neutrinos, axions have never been observed, but many theoretical physicists believe that in the very early universe, about a millionth of a second or less after the expansion began, axions could have been formed. Theory indicates that these particles are more massive than neutrinos, and that they should condense into galaxy-sized clumps. This implies, first of all, that galaxies are the fundamental entities, and that clusters represent a clumping together of galaxies. Secondly, it means that the fraction of missing mass around galaxies is about the same as in clusters of galaxies. Thus the hypothesis that the missing mass is in the form of axions seems more in accord with the observations. But several outstanding questions remain. Could superclusters form in an axion-dominated universe? At the other extreme, could dwarf galaxies form? Does yet another hypothetical particle, the photino, explain things better? The observations that can decide among the competing theories are still few and far between, and the implications of the theories remain to be fully explored.

Clearly, much more scientific detective work is needed to narrow the list of suspects in the mystery of the missing mass. In the meantime, if we rule out the possibility that the theory of gravity and dynamics is wrong, then our observations of galaxies and clusters of galaxies over the last decade have revealed a remarkable truth: the universe is not what it appears to be. Most of the matter of the universe is in an invisible form that cannot, as yet, be directly observed.

15

The Cosmic X-Ray Background

One of the first findings of x-ray astronomy was an unexpectedly strong background glow of x-radiation. The observations indicated that the x-ray background was of the same strength in all directions. This suggested that the radiation was not coming from some localized cloud, or the center of the galaxy, or even a few nearby galaxies, but from a distance so great that all local irregularities such as galaxies and clusters of galaxies merged into a smooth background, just as the lights of a distant city appear as a uniform glow. The implication was clear: the x-ray background carried information about the large-scale structure of the universe. This message was not lost on astronomers, and they moved quickly to interpret and use this information. As discussed in Chapter 6, one of the first applications of x-ray astronomical observations was to throw doubt on the steady-state theory of the universe. Now, twenty years later, the steady-state theory has all but vanished from the pages of scientific journals, but an understanding of the origin of the x-ray background is still just beyond our grasp.

How long ago was the x-ray background radiation produced? The current opinion is that it must have been about ten billion years ago. A combination of observation and theory strongly suggests that the universe as we know it began fifteen or twenty billion years ago as an explosion—the Big Bang as it is called—from a superhot, superdense state. When the universe was very young and very hot, it was filled with high-energy radiation. But this radiation shifted to much lower energies as the universe expanded and cooled, so it could not have been responsible for the x-ray background. A relic of this early state has been observed, but at much lower frequencies, in the form of the microwave background radiation discovered in the 1960s by Arno Penzias and Robert Wilson of Bell Laboratories. Astronomers believe the microwave background radiation was produced when the universe was about a million years old. The extreme uniformity of this radiation indicates that the universe was very smooth at that time, and that because of its rapid expansion it was much too cool to produce x-rays.

Sometime between a million and a billion years after the Big Bang, the universe made a dramatic transition from a smooth, featureless state to clumps of stars and galaxies. Why and how did this transition occur? The x-ray background must have been produced some-

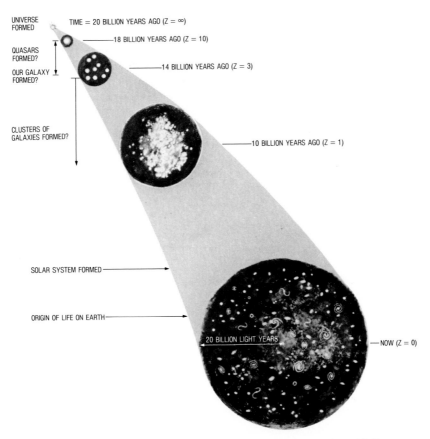

The evolution of the universe. (Sketch by B. Langer; courtesy M. Zombeck, Harvard-Smithsonian Center for Astrophysics)

time during this transitional period, so studies of this radiation tell us about one of the crucial phases in the history of the universe.

An analysis of the data from the *HEAO-1* satellite by Elihu Boldt and his colleagues at Goddard Space Flight Center shows that the spectrum of the x-ray background matches very closely that expected from a gas at a temperature of 400 million degrees. Using the theory of how a hot gas radiates x-rays, and assuming that the gas is spread throughout the space between galaxies, it is possible to compute the amount of intergalactic hot gas needed to explain the observed intensity of the x-ray background.

The calculations indicate that the total mass of the required gas would be greater than that of all the matter in all the galaxies, including the hidden mass. The mass is very nearly enough to close the universe, that is, to halt the present cosmic expansion and turn it into a collapse billions of years from now. If this is the situation, then the universe is finite, and the present phase will end in a cosmic fireball produced by the collapse. It is not necessarily unreasonable to believe there might be such a large amount of gas spread between the galaxies. It is quite possible that a large fraction of the gas that collapsed to form galaxies might have been left over in the form of hot intergalactic gas.

There are, however, serious problems for the theory that hot gas is the cause of the x-ray background radiation. First, as George Field and Stephen Perrenod of Harvard-Smithsonian have pointed out, the energy needed to heat the gas to the required 400 million degrees would be enormous. It is possible that this energy could be derived from the collapse of matter to form galaxies. Or maybe some other means could heat the gas, such as the energy given off by quasars and exploding galaxies. In this case, however, the energy output of quasars would have to be a hundred times greater than it is observed to be. Secondly, William Forman of Harvard-Smithsonian and his colleagues have shown that the hot background gas would evaporate most of the gas from clusters of galaxies. X-ray observations of clusters show that this has not happened; therefore the background gas must not have the required density and temperature.

Finally, and most convincingly, observations from the *Einstein* observatory indicate that distant individual sources such as quasars must make an appreciable contribution to the x-ray background. If the x-ray background is produced by a hot gas that fills all of space, then no matter how long an x-ray telescope looks at a particular region of the sky, the picture should not change: it should appear as a diffuse glow at all levels of sensitivity. But if the background, like the lights of a distant city, only appears to be smooth and is actually composed of many individual sources, then as we develop instruments that can detect fainter and fainter radiation, the discrete sources of radiation should begin to show up.

One of the primary goals of the *Einstein* observatory was to try to detect such sources. In deep surveys of blank fields of the sky — that is, regions empty of known radio, optical, or x-ray sources — dozens of new sources have been discovered. About two-thirds of these sources are quasars. An analysis of these results and the results from other *Einstein* observations indicates that at least 30 percent of the x-ray background comes from quasars and active galactic nuclei. We know from both x-ray and optical observations that quasars were more common in the past, reaching their peak about ten billion years ago. Since we do not know exactly how many quasars there were at each epoch in the past, or what their average brightness was, we cannot make an exact estimate of the contribution of the quasars to the x-ray background, but it is possible that they account for all of it. If they do, then we do not need to postulate hot gas to explain the x-ray background; and if the hot gas does not exist, the calculated mass of the universe is below the critical value. Unless it is filled with even more hidden matter than we have inferred from observations of galaxies and clusters, the universe will not be closed. Instead it is infinite, and will continue to expand forever.

A critical outstanding question is whether the observed spectrum of the x-ray background is consistent with radiation from distant quasars and active galaxies. In an effort to answer this and other questions, various scientists have used *HEAO-1* to measure the spectra of a number of nearby quasars and active galaxies. These spectra do not match the spectrum of the x-ray background. This is discouraging but not disastrous for those who believe the background is caused by quasars and active galaxies. The quasars that make the major contribution to the x-ray background are so distant and faint that their spectra cannot be measured with the *HEAO-1* instruments. It is possible that in the distant past the quasars and active galaxies produced a different spectrum of x-radiation from what they currently produce. Or perhaps they are not the source of the x-ray background after all. Perhaps the radiation comes from some pregalactic, prequasar stage of the universe.

Only with instruments more sensitive than those used on *HEAO-1* and *Einstein* will we be able to measure the number, intensity, and spectra of distant quasars and resolve the issue of what causes the x-ray background. When we finally understand the origin of the x-ray background, we will see more clearly "that immortal sea which brought us hither."

Coda

This survey gives some sense, we hope, of the travail, the excitement, and the delight that have attended the birth and growth of x-ray astronomy. From the nearest star to the edge of the universe, data from the x-ray universe challenge our theories and stretch our minds. It seems that the richness of the universe is virtually limitless, that any new voyage of discovery will be rewarded with new and wonderful insights. For this reason we cannot rest on the laurels of past successes. Each advance, be it Galileo's telescope, or the 200-inch telescope on Palomar Mountain, or the *Voyager* missions to Jupiter, or the *Einstein* x-ray observatory, to name but a few, answers old questions and raises new ones. The urge to know as much of our universe as possible drives us forward.

The need for new facilities is imperative in x-ray astronomy. Two satellites dedicated to x-ray astronomy, the Japanese *TENMA* and the European *EXOSAT*, were launched in 1983, and the German x-ray observatory *ROSAT* will be launched by the space shuttle in 1986. *EXOSAT* and *TENMA* are less powerful than the *Einstein* observatory, and their missions are limited to the study of relatively nearby sources. *ROSAT*, which will not be a permanent observatory, will be slightly more sensitive than the *Einstein* observatory, but will be limited to the study of x-rays below 1 kilovolt and will have very poor spectral resolution. It will nevertheless carry out a sensitive all-sky survey, whereas *Einstein* observed only 1 percent of the sky. Longer and more sensitive observations will be needed to extend our coverage of the sky and to study the spectra, time variability, polarization, and structure of the thousands of objects already discovered. To meet these needs, NASA has issued an An-

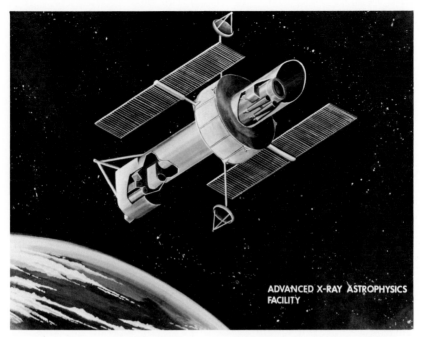

ADVANCED X-RAY ASTROPHYSICS
FACILITY

An artist's rendering of NASA's Advanced X-Ray Astrophysics Facility.
(NASA)

nouncement of Opportunity for a design study of an Advanced X-Ray Astrophysics Facility. The heart of this facility will be a 1.2-meter x-ray telescope. It is currently planned for launch in the mid-1990s. Some thirty years will have elapsed since Giacconi first proposed the launch of a 1.2-meter telescope. The concept of a free-flying x-ray observatory, maintainable in orbit and retrievable, has by now won wide support in the scientific community. The advanced x-ray facility will be operated as a national observatory. It will be a natural complement to similar facilities that already exist in optical and radio astronomy. In optical and radio astronomy, several complementary instruments exist for high resolution on the one hand and broad surveys on the other, such as the 200-inch reflector and the 48-inch Schmidt telescopes on Palomar Mountain and the Very Large Array and Very Long Baseline Interferometry in radio astronomy. Likewise, in x-ray astronomy instruments are under investigation — such as the Large Area Modular Array of Reflectors (LAMAR), an array of identical modules containing imaging telescopes and detectors — that would be especially useful in performing broad surveys of the sky and would otherwise complement the capabilities of the Advanced X-Ray Astrophysics Facility (AXAF).

AXAF should contribute to the solution of some of the outstanding problems in extragalactic astronomy. It will be possible to increase the sensitivity of the deep surveys by at least an order of magnitude. At these greater sensitivities, we will be seeing back in time to the epoch when galaxies formed and quasars were brightest. It should be possible to decide whether the cosmic x-ray background is due to quasars or to a different class of objects altogether, such as galaxies in the process of forming. The capability of AXAF will allow astronomers to study details of the x-ray jets in active galaxies, and to compile complete samples of active galaxies of differing optical and radio brightness. In this way it should be possible to clarify how quasars and active galaxies change in time and to understand the apparently fundamental nature of the nuclear emission of x-rays and how it relates to the source of power in the nuclei of active galaxies and quasars. With AXAF, astronomers will be able to study how clusters of galaxies evolve with time. This should help to resolve questions such as the origin of supergiant galaxies like M87 and the nature of the missing mass. The most valuable use of LAMAR will probably be the measurement of the spectra and time variations of the faintest extragalactic sources, for which AXAF is not well suited. Prime targets will be stellar coronas in our own galaxy, x-ray binaries in neighboring galaxies, active galaxies, and quasars.

As the Advanced X-Ray Astrophysics Facility and other projects come to fruition sometime in the next decade, we will open a permanent window on the x-ray universe. As the history of x-ray astronomy shows, such projects can be successful only if we plan and work with a combination of urgency and confidence in the future. The situation reminds us of the story of the man who wanted to plant a shade tree on his lawn.

"Plant it tomorrow," he told his gardener.

"But sir," the gardener protested, "it will be thirty years before that tree provides any shade."

"Then, by all means, plant it this afternoon."

Bibliographical Notes

Index

Bibliographical Notes

2. The Sensible World

The concept of a scientific paradigm has been popularized by Thomas Kuhn; see, for example, *The Structure of Scientific Revolutions*, 2nd ed. (Chicago: University of Chicago Press, 1970). See also Enrico Bellone, *A World on Paper: Studies on the Second Scientific Revolution* (Cambridge, Mass.: MIT Press, 1980).

3. Precursors

For accessible discussions of Roentgen's discovery of x-rays, see the *Life Science Library* book *Matter*, by Ralph Lapp and the Editors of *Life* (New York: Time, Inc., 1963); and George Gamow, *Biography of Physics* (New York: Harper, 1961). For more details on the life of Roentgen, see the article by G. Turner in the *Dictionary of Scientific Biography*, ed. C. Gillespie, 16 vols. (New York: Scribner's, 1970–1980), and references cited therein. For more technical details on x-rays, see the classic by A. Compton and S. Allison, *X-Rays in Theory and Experiment* (New York: Van Nostrand, 1935).

4. Pioneers

The primary references for this chapter are Herbert Friedman, "Rocket Astronomy," *Annals of the New York Academy of Science*, 198 (1972), 267, and a pamphlet by Friedman, *Reminiscences of Thirty Years of Space Research*, published by the Naval Research Laboratory as NRL Report 8113 (Washington, D.C., 1977). The early ionospheric research at NRL is described in Richard Hirsh's book on the history of x-ray astronomy, *Glimpsing an Invisible Universe* (Cambridge: Cambridge University Press, 1983). Hirsh also discusses the beginnings of rocket research, as do Willy Ley in

Rockets, Missiles and Men in Space (New York: Viking, 1968), Thomas de Gaiana in *Concise Encyclopedia of Astronautics* (Chicago: Follett, 1968), and Homer Newell in *Beyond the Atmosphere: Early Years of Space Science*, NASA History Series, NASA SP-4211 (Washington, D.C., 1980).

5. The Discovery of an X-Ray Star

The reactions to the launch of *Sputnik* and the establishment of NASA are discussed in Newell, *Beyond the Atmosphere*, Hirsh, *Invisible Universe*, and Ley, *Rockets, Missiles, and Men.* For the events leading up to the discovery of the first extrasolar source of x-rays, see Hirsh, *Invisible Universe;* R. Giacconi and H. Gursky, eds., *X-Ray Astronomy* (Dordrecht: D. Reidel, 1974), ch. 1; R. Giacconi et al., "The *Einstein* X-Ray Observatory," in *Telescopes for the 1980s*, ed. G. Burbidge and A. Hewitt (Palo Alto, Calif.: Annual Reviews, 1981); R. Giacconi, "1962–1972 (Up through *Uhuru*)," *Journal of the Washington Academy of Science 71*, 1 (1981); B. Rossi, "X-Ray Astronomy," *Daedalus* (fall 1976), p. 37. The quotations from Herbert Gursky are from S. Bleeker, "Riccardo Giacconi: X-Ray Astronomy Pioneer," *Star and Sky, 2* (Jan. 1980), 43. The paper reporting the discovery of Sco X-1 is R. Giacconi, H. Gursky, F. Paolini, and B. Rossi, "Evidence for X-Rays from Sources Outside the Solar System," *Physical Review Letters, 9* (1962), 435.

6. The Riddle of the X-Ray Stars

The basic references here are Giacconi and Gursky, eds., *X-Ray Astronomy;* Giacconi et al., "The Einstein X-Ray Observatory," Giacconi, "1962–1972," and Rossi, "X-Ray Astronomy." Other details and a sense of the state of development of x-ray astronomy and related topics in the 1960s can be obtained from articles written for the *Annual Review of Astronomy of Astrophysics:* for example, P. Morrison, "Extrasolar X-Ray Sources," vol. 5 (1967); R. Gould, "Intergalactic Matter," and R. Giacconi, H. Gursky, and L. Van Speybroeck, "Observational Techniques in X-Ray Astronomy," vol. 6 (1968); W. Hiltner and D. Mook, "Optical Observations of Extrasolar X-Ray Sources," A. Cameron, "Neutron Stars," and A. Hewish, "Pulsars," vol. 8 (1970); and B. Paczynski, "Evolutionary Processes in Close Binary Systems," vol. 9 (1971).

7. *Uhuru:* Neutron Stars and Black Holes

The basic references are the same as for Chapter 6. See also, in the *Annual Review*, G. Blumenthal and W. Tucker, "Compact X-Ray Sources," vol. 12 (1974); and D. Eardley and W. Press, "Astrophysical Processes Near Black Holes," vol. 13 (1975). And see H. Gursky and E. Van den Heuvel, "X-Ray-

Emitting Double Stars," and K. Thorne, "The Search for Black Holes," both in *New Frontiers of Astronomy: Readings from Scientific American* (San Francisco: Freeman, 1975).

8. The X-Ray Sky

Concise summaries of the developments in x-ray astronomy between *Uhuru* and *Einstein* are given in G. Clark, "X-Ray Astronomy from *Uhuru* to *HEAO-1*," *Journal of the Washington Academy of Science*, 71 (1981), 17; and in E. Boldt, "The High Energy Astronomy Observatory, *HEAO-1*," ibid., 71 (1981), 24. See also J. Culhane and P. Sanford, *X-Ray Astronomy* (New York: Scribner, 1981). For a detailed history of the HEAO program, see W. Tucker, *The Star Splitters* (Washington, D.C.: U.S. Government Printing Office, 1984). For details relating to the development of x-ray astronomy in the 1970s, see also the following *Annual Review* articles: P. Gorenstein and W. Tucker, "Soft X-Ray Sources," vol. 14 (1976); N. Bahcall, "Clusters of Galaxies," and H. Gursky and D. Schwartz, "Extra-galactic X-Ray Sources," vol. 15 (1977); J. Bahcall, "Masses of Neutron Stars and Black Holes in X-Ray Binaries," vol. 16 (1978); R. McCray and T. Snow, Jr., "The Violent Interstellar Medium," and G. Baym and C. Pethick, "Physics of Neutron Stars," vol. 17 (1979); H. Bradt and J. McClintock, "The Optical Counterparts of Compact Galactic X-Ray Sources," vol. 21 (1983). D. Pines gives a good review of the physics of collapsed stars in "Accreting Neutron Stars, Black Holes, and Degenerate Dwarf Stars," *Science*, 207 (1980), 597. The following *Scientific American* articles are also of interest: G. Clark, "X-Ray Stars in Globular Clusters" (Oct. 1977), and W. Lewin, "The Sources of Cosmic X-Ray Bursts" (May 1981).

9. A Telescope for X-Rays
10. *Einstein* into Orbit
11. First Light

The *Einstein* observatory is discussed in detail in Giacconi et al., *Telescopes for the 1980s*, and references given therein; less technical accounts are given in Tucker, *The Star Splitters*, and in R. Giacconi, "The *Einstein* X-Ray Observatory," *Scientific American* (Feb. 1980).

12. Stellar Coronas and Supernovas

X-ray emission from the solar corona is discussed in Giacconi and Gursky, eds., *X-Ray Astronomy*, and in an *Annual Review* article, G. Vaiana and R. Rosner, "Recent Advances in Coronal Physics," vol. 16 (1978). R. Noyes gives a less technical account in *The Sun, Our Star* (Cambridge, Mass.: Harvard University Press, 1983). X-ray emission from stars is reviewed in

P. Byrne and M. Rodono, *Activity in Red-Dwarf Stars* (Dordrecht: D. Reidel, 1983). V. Trimble reviews in detail the nature of supernova outbursts in "Supernova: The Event," *Reviews of Modern Physics 54* (1982), 1183. Research on supernova remnants is reviewed in an *Annual Review* article, S. Holt and R. McCray, "Spectra of Cosmic X-Ray Sources," vol. 20 (1982); and in *Supernova Remnants and Their X-Ray Emission*, ed. J. Danziger and P. Gorenstein (Dordrecht: D. Reidel, 1983).

13. Quasars and Active Galaxies

For an excellent nontechnical discussion of the discovery of quasars see T. Ferris, *The Red Limit* (New York: Bantam, 1979). A likewise excellent introduction to modern extragalactic research, including quasars, is J. Silk, *The Big Bang* (San Francisco: Freeman, 1980). For a review of current work on active galaxies and quasars, see the *Annual Review* article, B. Balik and T. Heckman, "Extranuclear Clues to the Origin and Evolution of Activity in Galaxies," vol. 20 (1982); and J. Hutchings, "QSOs: Recent Clues to Their Nature," *Publications of the Astronomical Society of the Pacific, 95* (1983), 799. The work of Schmidt and Green and that of Marshall et al. are published in *Astrophysical Journal, 269* (1983). For less technical treatments of active galaxies and quasars see the articles by A. Lightman and H. Tananbaum in *Revealing the Universe*, ed. J. Cornell and A. Lightman (Cambridge, Mass.: MIT Press, 1982), and the following *Scientific American* articles: T. Geballe, "The Central Parsec of the Galaxy" (July 1979), R. Blandford, M. Begelman, and M. Rees, "Cosmic Jets" (May 1982), and P. Osmer, "Quasars as a Probe of the Early Universe" (Feb. 1982).

14. Clusters of Galaxies and the Missing Mass

Einstein observations of clusters of galaxies are reviewed in an *Annual Review* article, W. Forman and C. Jones, "X-Ray Imaging Observations of Clusters of Galaxies," vol. 20 (1982). The hidden mass is discussed in G. Blumenthal et al., "Formation of Galaxies and Large-Scale Structure with Cold Dark Matter," *Nature* (in press), and in an *Annual Review* article, S. Faber and J. Gallagher, "Masses and Mass-to-Light Ratios of Galaxies," vol. 17 (1979). Less technical discussions are given in G. Field's article in Cornell and Lightman, eds., *Revealing the Universe*, and in the following *Scientific American* articles: P. Gorenstein and W. Tucker, "Rich Clusters of Galaxies" (Nov. 1978), and V. Rubin, "Dark Matter in Spiral Galaxies" (June 1983).

15. The Cosmic X-Ray Background

For a discussion of the early work on the x-ray background see the *Annual Review* article by J. Silk, "Diffuse X and Gamma Radiation," vol. 11

(1973); and Boldt, "The High Energy Astronomy Observatory, *HEAO-1*." The x-ray spectra of quasars and active galaxies are discussed in Holt and McCray, "Spectra of Cosmic X-Ray Sources," and the contribution of active galaxies and quasars to the background is discussed in Giacconi et al., *Astrophysical Journal* 234 (1979): L1, and by S. Murray in *X-Ray Astronomy with the Einstein Satellite*, ed. R. Giacconi (Dordrecht: D. Reidel, 1981). For a less technical discussion see B. Margon, "The Origin of the Cosmic X-ray Background," *Scientific American* (Jan. 1983).

Index